SHARKS & RAYS *of the* ARABIAN/PERSIAN GULF

Dareen K. Almojil

Alec B.M. Moore

William T. White

First Published in Great Britain in 2015

British Library Cataloguing in Publication Data
A catalogue record for this book is available
from the British Library

Published by
MBG (INT) Ltd
282 Harrow Road, London W2 5ES, UK
Tel: +44 (0) 2072899000 Fax: +44 (0) 2072899009
Email: info@mbgprint.co.uk   www.mbgprint.co.uk

ISBN

978-0-9930427-2-0

Printed by

MBG (INT) Ltd

Cover design by Mohammed Sharaf
Illustrations by Georgina Davis & Lindsay Marshall
Book design by William White, Alec Moore & Dareen Almojil
Layout by William White

Foreword

This book, *Sharks and Rays of the Arabian/Persian Gulf,* is an important contribution to the literature of elasmobranchs of the Arabian Gulf, documenting the different species that inhabit the Gulf waters. It offers amazing insight for every elasmobranch, and all those interested in the marine life of the Gulf. This is the first comprehensive resource book on these most amazing marine animals, which have not been thoroughly studied. The authors' valuable efforts in compiling the scientific information and in documenting the diversity of sharks and rays of the Gulf waters are highly appreciated. The book will be a valuable resource reference for ichthyologists, students, biodiversity experts and ecologists, as well as those seeking a greater understanding of these fascinating creatures.

Dr. Faiza Yousef Al-Yamani

Executive Director of the Environment and Life Sciences Research Center
Kuwait Institute for Scientific Research
Kuwait

Table of contents

Introduction

Geography

The Gulf[1] is a shallow inlet of the Arabian Sea (itself part of the northwestern Indian Ocean) that lies west of the Straits of Hormuz between the Arabian Peninsula and the Islamic Republic of Iran. The Gulf is semi-enclosed, with a single narrow entry at the Straits of Hormuz to the adjacent and much deeper Gulf of Oman, which reaches depths of over 3,000 m. Eight countries have a coastline on the Gulf: Iran, Iraq, Kuwait, Saudi Arabia, Bahrain, Qatar, United Arab Emirates (UAE), and, on the Musandam Peninsula, a small exclave of the Sultanate of Oman (Figure 1).

Physical environment

The Gulf is a shallow and young sea. Only about 20,000 years ago, sea levels worldwide were much lower due to water being locked up during the last ice age and the current Gulf was just a river valley. About 14–17,000 years ago, the Gulf was flooded as global temperatures increased, sea levels rose, and it was only around 7,000 years ago that the sea level and coastlines we see today became stable. The average depth of the Gulf is only 35 m, and large areas of the southern Gulf are less than 20 m in depth. Only in the northeastern most part near the Straits of Hormuz depths do increase to a maximum of around 100 m.

Surrounded by arid landmasses, the shallow and semi-enclosed Gulf is an exceptionally harsh environment for marine organisms to live in. Water temperatures can fluctuate dramatically between seasons (11–36°C), but are normally high, causing evaporation and, in turn, relatively high salinity compared to other seas worldwide. This high salinity is exacerbated by the fact that there is minimal freshwater input into the Gulf to counteract this, and there is very limited exchange with lower-salinity oceanic water through the narrow Straits of Hormuz.

The unique elasmobranch fauna of the Gulf

The physical environment of the Gulf, most notably its shallow depth, significantly influences its marine biodiversity. The diversity of sharks and rays that are normally found in the Gulf appears to be different from the adjacent deeper waters, such as the

---

[1] **Arabian Gulf or Persian Gulf?**
The 'correct' name for the Gulf remains contentious, and could itself form the subject of an entire book. This book is intended for scientific use and wishes to avoid offending any sensitivities as far as possible, while simultaneously accepting that this is both inevitable and unavoidable. As a best compromise, we therefore use the full term 'Arabian/Persian Gulf' (alphabetical order), and, as a short form, use the now widely-used term 'Gulf'. The Gulf of Oman (also known as the Sea of Oman) is always referred to by its full name.

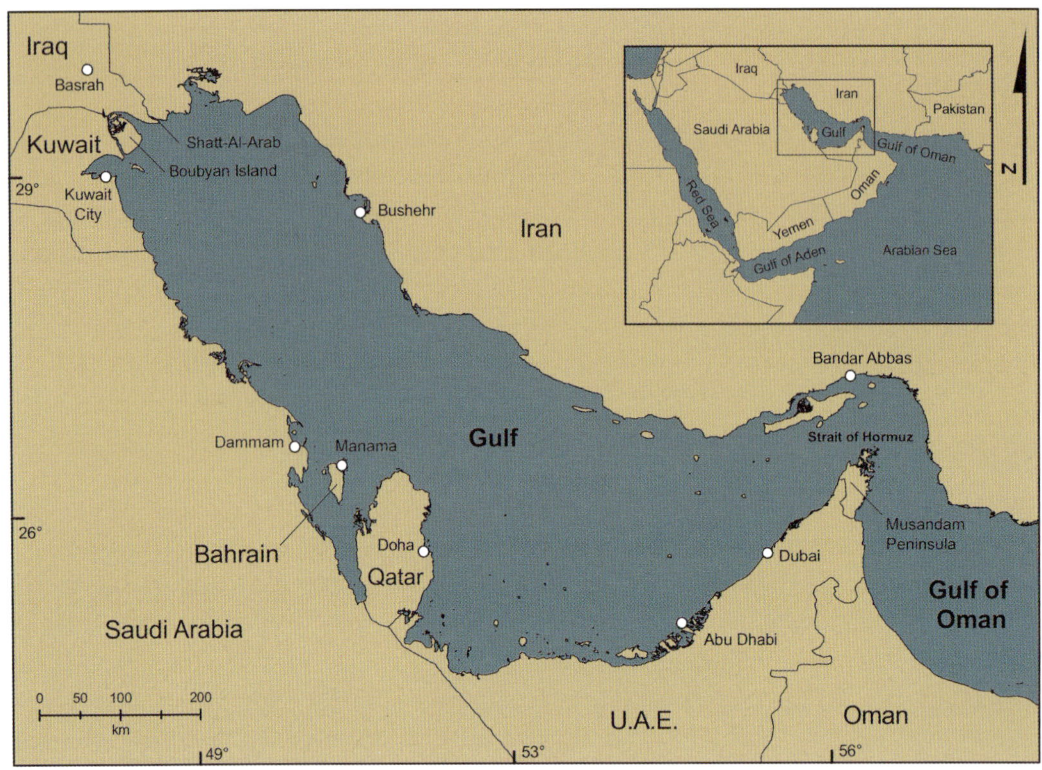

Figure 1 Map of the Gulf region showing all countries and major cities.

Gulf of Oman. With the possible exception of the deeper waters in the easternmost Gulf, the following groups of sharks and their relatives, normally associated with deeper waters, are not known to occur in the Gulf: 1) large oceanic pelagic species (e.g. Oceanic Whitetip *Carcharhinus longimanus*); 2) deepwater species (e.g. Bigeye Houndshark *Iago omanensis* and Bramble Shark *Echinorhinus brucus*); and 3) the chimaeras or ratfishes (e.g. *Chimaera* and *Neoharriotta*).

Conversely, the Gulf has its own unique species. The Arabian Banded Whipray *Himantura randalli* might only be found in the Gulf, and the rare smoothtooth blacktip shark *Carcharhinus leiodon* is known only from the Gulf and the Arabian Sea. It is likely that further research will reveal how unique and important the elasmobranch fauna of the Gulf is.

Key habitats

The majority of the Gulf is characterised by water less than 40 m depth with a soft and relatively featureless seabed of mud and/or sand. This typical habitat probably supports very common and widespread species such as the Milk Shark *Rhizoprionodon acutus*, Whitecheek Shark *Carcharhinus dussumieri* and Spottail Shark *Carcharhinus sorrah*. Species found in the upper, surface layers of water in the Gulf include active predators such as the Spinner Shark *Carcharhinus brevipinna*, the filter-feeding Whale Shark *Rhincodon typus*, and devilrays *Mobula* spp.

In addition, the Gulf also provides a number of other, more distinctive, habitats including seagrass beds, estuaries, mudflats, mangroves and coral reefs. Seagrass or macroalgae (such as *Sargassum*) dominates large areas of shallow waters, particularly in the southern central Gulf around Bahrain, Qatar and Abu Dhabi. Large seagrass meadows provide food for large herbivores such as dugong and green turtle, which in turn may attract one of their main predators, the Tiger Shark *Galeocerdo cuvier*.

The only major estuary, the Shatt al-Arab, flows into the northwestern Gulf, between Iraq and Iran. This estuary combines the flow from Tigris and Euphrates rivers, both of which originate in distant Turkey, and the Karun River from Iran. There are a few other much smaller estuaries along the Iranian coastline, but none are found along the Arabian coast. The majority of elasmobranch species worldwide are restricted to marine or slightly brackish water, but a few species can tolerate fresh water of very low salinity. A notable example is the Bull Shark *Carcharhinus leucas*, which is well known for its ability to penetrate fresh water and may do so to give birth to its pups in a more sheltered environment. Bull sharks have been recorded as far upstream as Baghdad, and are believed to be responsible for biting humans in rivers in a number of incidents in Iraq and Iran. It is likely that one or more species of sawfish (*Anoxypristis* and *Pristis*) also used local estuaries and may have penetrated into freshwater, but it is likely that these critically endangered animals are now essentially extinct in the Gulf.

The Gulf's estuarine environments are associated with intertidal mudflats such as those in the north of Kuwait. These habitats can be highly productive, with large abundances of prey such as oyster reefs, crabs and mudskippers. Species such as the Longheaded Eagle Ray *Aetobatus flagellum* appear to be restricted to these estuarine-influenced, turbid waters, and large numbers of guitarfish (*Glaucostegus* spp.) have been observed feeding in extremely shallow waters on an incoming tide.

Due to natural environmental conditions, in addition to man-made degradation, mangroves are not widespread in the Gulf, with the only significant stands on the coasts of Iran and the UAE. Mangroves play an extremely important role in coastal ecosystems, for example through stabilising sediments and providing nursery areas for commercial fish and shellfish species. Elsewhere in the tropics, mangroves can be essential nursery habitats for elasmobranchs such as Lemon Sharks *Negaprion acutidens* and sawfish. There is only one mangrove species (*Avicennia marina*) that occurs naturally in the main part of the Gulf. To date, there has been no research on the elasmobranchs of the Gulf's mangrove ecosystems, but it is likely that these habitats are of significant interest in terms of diversity, abundance, and potential importance as nursery areas.

Coral reefs of the Gulf display less diversity when compared to other Western Indian Ocean locations; the Gulf's extreme and highly fluctuating environmental conditions plays a major role in restricting coral reefs from flourishing. Rocky habitats are also limited in extent (such as to the coast of Saudi Arabia and the Musandam Peninsula), therefore the shark and ray species that one might expect on a tropical reef (such as Whitetip Reef Sharks *Triaenodon obesus*) are not well documented in the Gulf and may be rare or possibly absent. A few reef-associated species have been recorded (such as the Blacktip Reef Shark *Carcharhinus melanopterus* and Grey Reef Shark *Carcharhinus amblyrhynchos*), but these appear to be relatively limited in their distribution and abundance. The Arabian Carpetshark *Chiloscyllium arabicum*, however, is commonly observed by divers.

How to use this book

The aim of this book is to document the species of sharks and rays that occur in the Gulf and those which might be seen in Gulf fish markets. The species profiles in this book are divided into two sections: those confirmed as occurring in the Gulf and those possibly occurring in the Gulf. Those listed in the latter section are i) species that are likely to occur in the Gulf but have not been reported yet, ii) species that occur in adjacent waters (e.g. Gulf of Oman), and could occur in the Gulf, and iii) species caught outside of the Gulf but transported to Gulf fish markets (e.g. Dubai) for sale.

An illustrated glossary is provided which highlights the anatomical terms and various terminology used throughout the book. A key is provided so that a specimen can be systematically identified to its family and genus if required. The families or genera marked with an asterisk(*) in the key are **possibly occurring** species (pp. 134–160).

The shark and ray species confirmed as occurring in the Gulf are provided first in this book and each species consists of two facing pages. The families of sharks and rays are arranged in taxonomic order beginning with the Hemiscylliidae through to the Mobulidae. Each order of sharks or rays is colour coded (on the right hand margin of the page spread) and within each family, species are in alphabetical order. Following the occurring shark and ray species, the possibly occurring shark and ray species are provided. These are also listed in taxonomic order of families and by alphabetical order within the families, but they are not colour coded by order but as "Possibly occurring species". Each possibly occurring species consists of a single page.

Species profiles

For each species of shark and ray occurring in the Gulf, we provide information on its size, key features for identification, worldwide distribution, Gulf occurrence, habitat and biology, conservation status, remarks, specific references used, and details of images used. For the possibly occurring species, similar information is provided but in a reduced format.

Common & scientific names

The English common names generally follow those adopted by Ebert *et al.* (2013) and Last & Stevens (2009). The scientific name of each species consists of two parts; genus and species names. The scientific name is followed by the name of the author(s) who named it, and the year in which it was named. Parentheses around the author(s) and date indicate that the author originally placed the species in a different genus. New species, or species of uncertain identity, are referred to by a generic name and 'sp.' (e.g. *Rhinobatos* sp.). Alternatively, 'cf. ' is placed between the generic and species names if the species is similar to, but possibly different from, the named species.

Images and illustrations

Where possible, the primary image used to depict each species in this book belongs to a specimen collected from the Gulf. When a Gulf image was not available, a suitable image from an adjacent region (e.g. Gulf of Oman, Red Sea or Indian Ocean) was used. The location details of the images used are provided in the image details section for each species. The main exception is for the Ganges Shark (*Glyphis gangeticus*) for which a drawing of one of the types was used, as the only available images were of poor quality museum specimens.

Additional images were used, where possible, to highlight features indicative of a particular species that are not apparent on the primary image. In some cases, line drawings were used to better depict these features. Line drawings used in this book were drawn by Georgina Davis and Lindsay Marshall.

Size

Sizes provided for each species in this book are either total length (TL) or disc width (DW). In the case of sharks, shark-like rays (e.g. guitarfishes and sawfishes) and the electric rays, TL is measured as a straight line from the tip of the snout to the tip of the extended upper caudal-fin lobe. For stingrays, eagle rays, cownose rays and devilrays, DW is measured across the disc at its widest point. In addition to the maximum size, the size at sexual maturity for both sexes and the size at birth are given for each species when known. Since size data for local species is lacking in most cases and size characteristics may vary from one region to another, it is important to note that some of the provided size information in this book are gathered based on data from other regions. For some species where data is available from the Gulf, this is specified in parentheses.

Key features

This section lists the diagnostic features by which a species can most easily be identified in the field. These features include body shape, colour patterns, shape of fins and their positions, and tooth shape. Each character is ordered numerically and the corresponding number is positioned on the relevant image to illustrate the feature.

Worldwide distribution

The known global geographic distribution for each species is provided in this section, e.g. Indo-West Pacific, from South Africa to northern Australia.

Gulf occurrence

This section provides a brief account of the known occurrence for each species in the Gulf based on species-specific publications from the region and new records and, if known, its relative abundance. Most species are probably more widely distributed in the Gulf than listed here, but there is typically limited information due to the paucity of data on sharks and rays in this region. The information in this section will act as a baseline for future research. Since there are no confirmed records from the Gulf of the possibly occurring sharks and rays listed in this book, this section is omitted for these species treatments.

Habitat and biology

This section provides three core pieces of information: habitat, feeding and reproduction. The basic habitat information provided for each species includes the typical habitat occupied (e.g. pelagic and oceanic, demersal on continental shelf, coastal), any specific habitat related behaviour and known depth range. The feeding information provided for each species lists the main prey items consumed. The reproduction section for each species includes the mode of reproduction, and the litter size and gestation period if known. The limited local research on sharks and rays again hinder our understanding on the biology and ecology of local species, thus the provided information is mainly based on studies in other regions.

Conservation status

This section lists the current global red list assessment category for each species from The *IUCN Red List Threatened Species*™ listed in this book. We also include the year of assessment.

Remarks

This section contains information that is considered to be potentially relevant to the reader accessory to the sections listed above. This includes information such as taxonomic concerns, specific behavioral attributes, similar species, etc. For the possibly occurring sharks and rays, this section explains the reason for its possible occurrence in the Gulf.

References

Within the species treatments, specific references(s) that were used to collect information that are species and/or geographically specific are provided.

Image details

This section provides the location, size and sex, and date of the specimen(s) in the images.

References

References used throughout the book are cited in full. As well as the specific references listed in species profiles, general references used throughout this book are also provided; these include: Carpenter & Niem (1999), White *et al.* (2006), Last & Stevens (2009), Last *et al.* (2010), and Ebert *et al.* (2013).

Image copyright information

In this section, the copyright holders of the images used for each species are provided. For each copyright holder, a list of the species with the image type (e.g. lateral, ventral head) in parentheses is provided.

Indexes

Indexes of scientific names and common names are listed on pages 174–178.

Glossary

barbel – a tentacle-like sensory structure on the head.

benthic – living on the seabed.

blotch – a patch that is different in colour to adjacent areas.

brackish – waters with a salinity between freshwater and saltwater, as in river estuaries.

bycatch – non-target catch components caught by artisanal and commercial fishers.

caudal peduncle – the narrow posterior part of the body connecting the caudal fin.

cephalic lobe – a flat roundish projection on the forehead of some rays.

cephalopod – a group of animals including cuttlefishes, nautili, squids and octopi.

cetacean – a group of aquatic mammals that includes whales, dolphins and porpoises.

circumglobal – distributed around the world within a certain latitudinal range.

claspers –modified parts of pelvic fins which males possess that are used to transfer sperm to the female during mating.

cloaca – a common opening for digestive, urinary and reproductive tracts.

common name – the informal given name for an animal, which may vary across the geographic range for a species.

compressed – flattened laterally from side to side.

concave – curved inwards (opposite of convex).

continental shelf – the shelf-like part of the seabed adjacent to the coast to a depth of about 200 m.

continental slope – the typically steep, slope-like part of seabed bordering the continental shelf to a depth of about 2000 m.

convex – curving outwards (opposite of concave).

cosmopolitan – having a worldwide distribution.

crescentic – shaped like a new moon (roughly C-shaped).

crustaceans – a group of invertebrate animals including crabs, shrimps, prawns, lobsters and crayfish.

cusp – a projection on a tooth.

cusplet – a small cusp.

demersal – living on or near the seabed.

denticle – a tooth-like structure; scale of a shark or ray.

depressed – flattened ventrally; from top to bottom sides of the body.

depth – height of body or head from top to bottom; also distance from sea surface to the bottom.

disc – the combined head, trunk and enlarged pectoral fins of those cartilaginous fishes with depressed bodies, e.g. stingrays.

dorsal – relating to upper part or surface of back.

dorsolateral – positioned or orientated between dorsal and lateral surfaces.

dusky – darkish to greyish in colour.

elasmobranch – a group of fishes comprising the sharks and rays.

electric organ – an organ that produces an electrical discharge.

elongate – extended in length in relation to width.

endemic – native and restricted to an area.

epipelagic – the upper body of water the extends from the surface to about 200 m.

estuarine – an organism that lives or found in estuaries.

falcate – curved like a sickle.

family – a classification term used for grouping organisms, containing one or more closely-related genera.

fauna – the communities of animals in an area.

filter feeding – a feeding strategy that uses special structures (ex. gill rakers) to sieve food particles from the water.

genus – a classification term used for grouping organisms, containing one or more closely-related species.

gillnet – a net used to tangle fishes.

gill opening – an opening (usually slit-like in sharks and rays) on head that connects the gill chamber to the exterior.

habitat – the environment in which an organism lives.

hammer-shaped – shaped with paired lateral expansions, like the head of a mallet.

head – specialised anterior part of an animal on which the mouth and major sensory organs are located; part other than body and tail.

histotrophy – form of embryonic nutrition where the developing embryos receive a lipid-rich histotroph, or uterine milk , usually delivered through extensions of the uterine wall called trophonemata.

hyomandibular pores – line of enlarged pores on both sides of the mouth corners.

interdorsal – area between the first and second dorsal fins.

interdorsal ridge – ridge of skin between dorsal fins.

internarial space – space between the nostrils.

internasal flap – fleshy flap extending between nostrils, sometimes partly covering the mouth.

interorbital space – area on top of head between eyes.

jaws – part of mouth supporting teeth.

juvenile – young fish, not yet sexually mature.

keel – a fleshy ridge.

labial furrow – shallow groove sometimes present at corners of mouth.

lateral – referring to the sides.

lateral ridges – fleshy expansions on sides of body.

longitudinal – lengthwise (opposite of transverse).

longline – a fishing line with a number of baited hooks.

lunate – shaped like a crescent moon.

margin – edge or rim.

median – relating to the middle of an object.

mesopelagic – living in open ocean at depths between 200 and 1000 m.

nape – region of head above and behind eyes.

nictitating eyelid – a transparent, moveable membrane or inner eyelid that protects eye.

nostril – external opening of the nasal organs.

obtuse – broadly rounded or having a blunt end.

oceanic – living in the open ocean.

ocellus (pl. ocelli) – an eye-like marking.

oophagy – method of matrotrophic embryonic nutrition in viviparous species where the embryo feeds on unfertilised eggs in the uterus.

orbit – a cavity in the skull that houses the eyeball.

oviparous – producing eggs that hatch after being deposited from the body of a pregnant female.

pelagic – free-swimming in the seas, oceans or open water and not associated with the bottom.

plain – uniformly coloured, without a contrasting colour pattern.

placental – method of matrotrophic embryonic nutrition in viviparous species where nutrients are transferred across the mother's uterine epithelium, which is intimately connected with foetal tissue (placenta).

plankton – small microscopic organisms that drift or float in open water.

population – a biological unit that represents the individuals of a species living in a certain area.

pore – a small opening in the skin that has secretory or sensory functions.

posterior – relating to hind of or rear end of an object.

precaudal pit – a transverse or longitudinal notch on caudal peduncle just anterior to origin of caudal fin in some sharks.

quadrangular – shaped with four distinct edges or margins.

reticulated –a network arrangement.

rhomboidal – diamond-shaped.

rostral cartilage – a gristly structure supporting the snout.

rostral teeth – tooth-like projections on the sides of the snout of sawfishes and sawsharks.

rostrum (adj. rostral) – a projecting snout.

salinity – the concentration of salt in water.

school – a close aggregation of fish swimming in association with each other.

scientific name – the formal binomial name of an organism consisting of the genus and species names; only one valid scientific name exists per species.

serrate – saw-like.

snout – part of head in front of eyes.

species – actually or potentially inter-breeding populations that are reproductively isolated from other populations.

spiracle – a respiratory opening behind the eye in sharks and rays.

spiracular fold – fold of skin present on the hind margin of the spiracular opening in some ray species.

spot – a regularly shaped or rounded area of a colour different to adjacent areas.

stinging spine – large, serrated bony structure on the tail of some rays.

subequal – nearly equal.

substrate – the substance forming the bottom of the sea or ocean.

subterminal – positioned near but not at end of an object.

symphysis (adj. symphysial) – relating to where the two halves of either a lower or upper jaw join.

synonym (adj. synonymous) – each of two or more scientific names of the same rank used to denote the same taxon.

tail – part of fish between cloaca and origin of the caudal fin.

taxonomy – the science of classification of organisms.

terminal – located at or forming the end of something.

thorn – large denticles on surface of a ray or skate.

tip – the extremity of a part of a fish.

tooth rows – horizontal rows of teeth in the jaws.

total length – longest length of a fish, from snout tip to upper caudal tip or tail tip.

transverse – directed crosswise, across width (opposite of longitudinal).

trawl net – fishing net which is dragged behind a boat.

trunk – part of an elasmobranch between head and tail; between fifth gill slit and cloaca.

ventral – relating to the lower part or surface.

vermiculations – wavelike or wormlike decoration pattern.

vertebrate – animal having a vertebral column or backbone.

viviparous – producing live young from within the body of the parent female.

Structural features of sharks

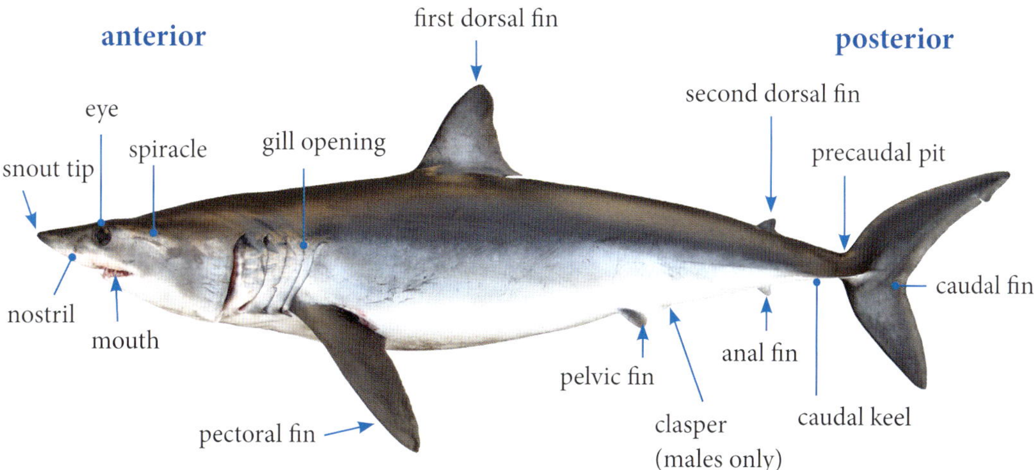

Shark dorsal fin

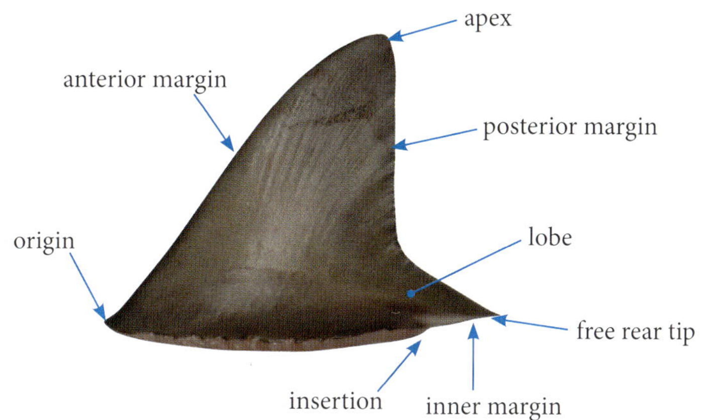

Shark ventral head

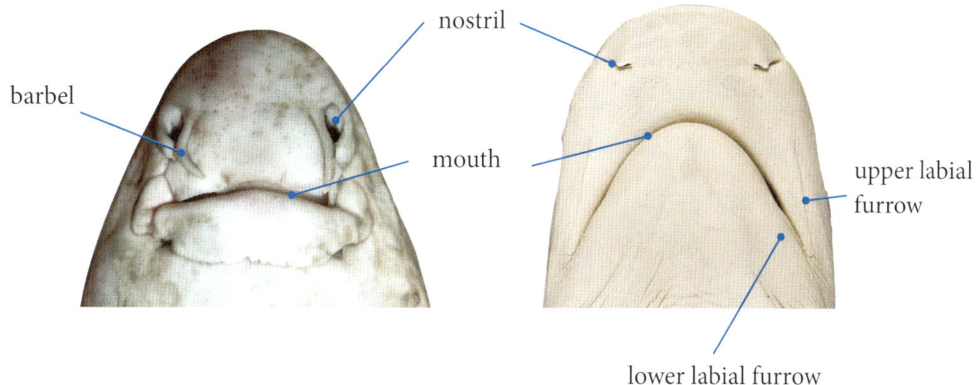

Shark caudal fin

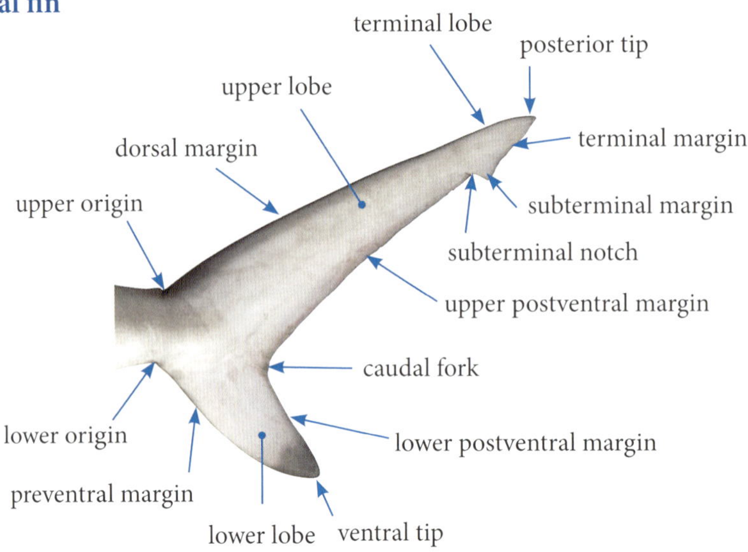

Structural features of rays

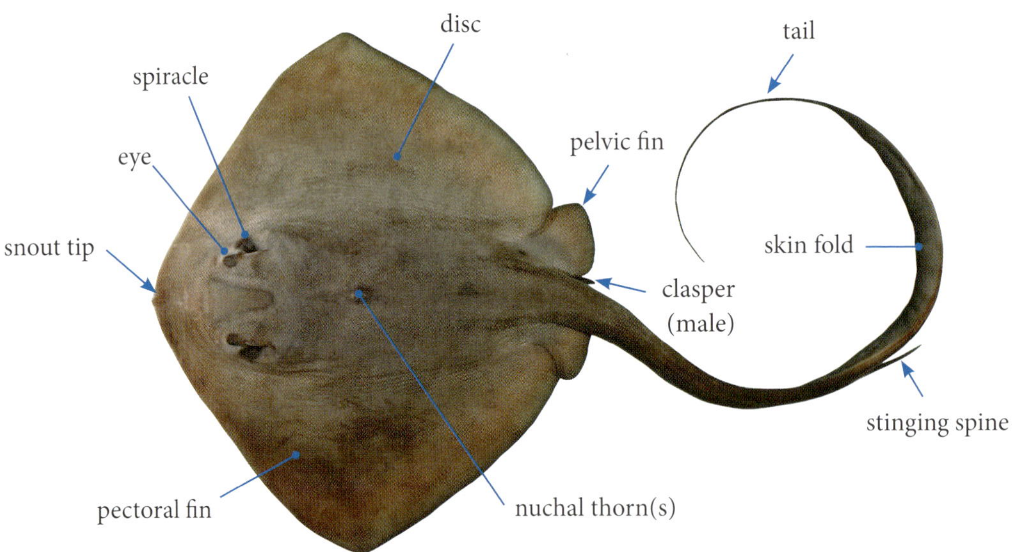

Oronasal (nostrils and mouth) region of rays

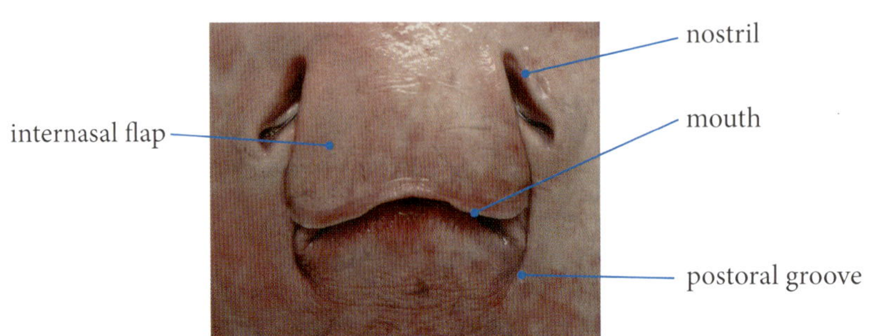

Key to families and genera

1. Gill openings on sides of head 2

    Gill openings on undersurface of head 12

2. Head laterally expanded, hammer-shaped ....
    .... Family Sphyrnidae, Hammerhead Sharks

    Head very broad, narrow and wing-like; width of
    head about half of total length ....
    .... *Eusphyra** (fig. 1; p. 148)

    Head not as broad or wing-like; width of head less
    than 40% of total length ...... *Sphyrna* (fig. 2; p. 82)

    Head not hammer-shaped 3

3. Upper caudal-fin lobe very long; equal to or
    more than half total length; body plain ....
    .... Family Alopiidae, Thresher Sharks
    fig. 3; p. 134

    Upper caudal-fin lobe much less than half total
    length (caudal fin also long in *Stegostoma*, but
    body spotted and/or banded) 4

4. Whole mouth forward of front margin of eye
    (fig. 4) 5

    Mouth partly beneath or behind front margin of
    eye (fig. 5) 8

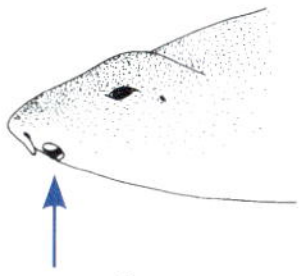

5. Mouth very broad and at front of head (terminal);
    caudal fin with a well-developed lower lobe ......
    .... Family Rhincodontidae, Whale Sharks
    fig. 6; p. 32

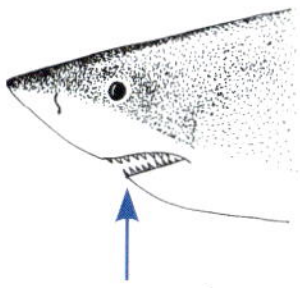

    Mouth narrower and not at front of head
    (subterminal); caudal fin without an obvious
    lower lobe 6

6. Outer margin of nostril without a fleshy lobe or
    groove (fig. 7) ....
    .. Family Hemiscylliidae, Longtail Carpetsharks
    fig. 9; p. 26

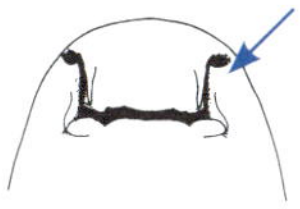

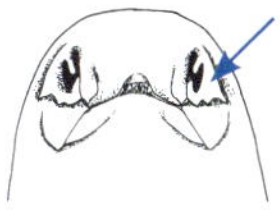

    Outer margin of nostril with a fleshy lobe and
    groove (fig. 8) 7

fig. 1

fig. 2

fig. 3

fig. 4

fig. 5

fig. 6

fig. 7

fig. 8

fig. 9

7. Caudal fin very long, almost equal to trunk length; strong ridges present on side of body ....
    ....Family Stegostomatidae, Zebra Sharks
    fig. 10; p. 30

    Caudal fin shorter, less than half trunk length; no ridges on sides of body ....
    .... Family Ginglymostomatidae, Nurse Sharks
    fig. 11; p. 28

fig. 10

fig. 11

8. Caudal fin upper and lower lobes of almost equal length; a strong keel on either side of caudal peduncle ....
    .... Family Lamnidae*, Mackerel Sharks

    Upper teeth broadly triangular and flattened, with strongly serrated edges...
    .... *Carcharodon** (fig. 12; p. 136)

    Upper teeth narrow and not flattened, with smooth edges .... *Isurus** (fig. 13; p. 137)

fig. 12

fig. 13

    Caudal fin upper lobe much longer than lower lobe; no or only a low keel on each side of caudal peduncle 9

9. Eyelid fixed, not capable of closing over eye ......
    ....Family Odontaspididae
    fig. 14; p. 34

    Eyelid nictitating (capable of closing over eye)
    10

fig. 14

10. Precaudal pits absent; leading edge of upper lobe of caudal fin smooth ....
    .... Family Triakidae, Hound Sharks

    Teeth in jaws blade-like (for cutting) ....
    .... *Iago** (fig. 15; p. 139)

    Teeth in jaws pavement-like (for crushing) ....
    .... *Mustelus* (fig. 16; p. 36)

fig. 15

fig. 16

    Precaudal pits present; leading edge of upper lobe of caudal fin typically rippled 11

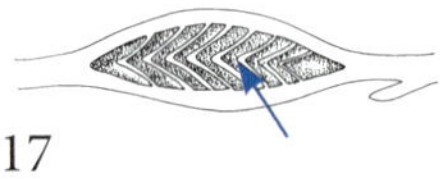

11. Small spiracles present; intestine with spiral valves (fig. 17) ....
    .... Family Hemigaleidae, Weasel Sharks

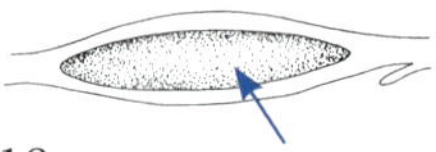

fig. 17

fig. 18

a. Lower teeth near symphysis with short, straight or weakly hooked cusps (fig. 19) that are mostly concealed when mouth closed; gill slits small, less than twice length of eye .... *Paragaleus* (fig. 21; p. 42)

Lower teeth near symphysis with long, strongly hooked cusps (fig. 20) that prominently protrude from mouth when closed; gill slits large, more than twice length of eye .... b

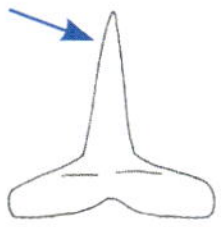

b. Snout obtusely wedge-shaped in dorsoventral view; fins not falcate, posterior margins of pelvic and pectoral fins straight or slightly concave .... *Chaenogaleus* (fig. 22; p. 38)

Snout bluntly rounded in dorsoventral view; fins strongly falcate, posterior margins of pelvic and pectoral fins deeply concave .... *Hemipristis* (fig. 23; p. 40)

Spiracles absent (except in the tiger shark *Galeocerdo cuvier*); intestine with a scroll valve (fig. 18) .... Family Carcharhinidae, Whaler Sharks

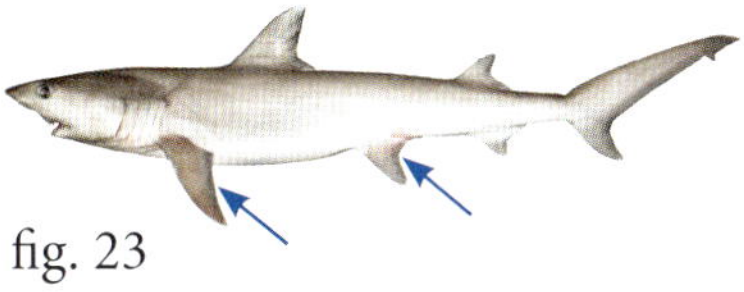

a. Caudal peduncle with a low keel on each side; small spiracles present; upper labial furrows very long, reaching forward to front of eyes .... *Galeocerdo* (fig. 24; p. 72)

Caudal peduncle without lateral keels (very weak ones present in *Prionace*); spiracles absent; upper labial furrows short (never reaching forward to in front of eyes) .... b

b. Second dorsal fin half or greater height of first dorsal fin c

Second dorsal fin less than half height of first dorsal fin .... e

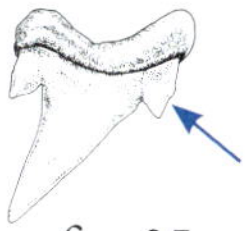

c. First dorsal and upper caudal fin with distinct white tips; teeth with a single cusp and one or more large lateral cusplets (fig. 25) .... *Triaenodon** (fig. 26; p. 147)

First dorsal and upper caudal fin without white tips; teeth with a single cusp, no lateral cusplets .... d

d. Second dorsal fin almost same height as first dorsal fin; upper precaudal pit deep and cresent-shaped .... *Negaprion* (fig. 27; p. 76)

fig. 19   fig. 20

fig. 21

fig. 22

fig. 23

fig. 24

fig. 25

fig. 26

fig. 27

Second dorsal fin about half height of first dorsal fin; upper precaudal pit a shallow, longitudinal depression .... *Glyphis** (fig. 28; p. 144)

e.   Head long, greatly flattened and trowel-shaped; pectoral fins broadly triangular; first dorsal fin well back on body (free rear tip almost over mid-base of pelvic fins) .... *Scoliodon** (Fig. 29; p. 146)

Head varying from conical to slightly flattened; pectoral fins narrower; first dorsal fin further forward, free rear tip over or slightly anterior to pelvic-fin origin .... f

f.   Second dorsal-fin origin well behind anal-fin origin (usually over or just anterior to anal-fin insertion); posterior margin of anal fin nearly straight or shallowly concave; ridges in front of anal fin very long (subequal to anal-fin base length) .... g

Second dorsal-fin origin usually about level with anal-fin origin (sometimes more posterior but always well anterior of anal-fin insertion); posterior margin of anal fin deeply concave or notched; ridges in front of anal fin short (about half anal-fin base or less) .... h

g.   Notch present at posterior edge of eye (fig. 30); first dorsal-fin origin well behind adpressed pectoral-fin free rear tips .... *Loxodon* (fig. 31; p. 74)

Posterior edge of eye without a notch; first dorsal-fin origin over or only just behind adpressed pectoral-fin free rear tips .... *Rhizoprionodon* (fig. 32; p. 78)

h.   First dorsal-fin base much closer to pelvic-than pectoral-fin bases; colour brilliant dark blue above; a low, weak keel present on each side of caudal peduncle .... *Prionace** (fig. 33; p. 145)

First dorsal-fin base centred between pectoral and pelvic-fin bases; colour light to dark greyish, greyish brown or brownish above; no keels on caudal peduncle .... *Carcharhinus* (fig. 34; p. 44)

fig. 28

fig. 29

fig. 30

fig. 31

fig. 32

fig. 33

fig. 34

12. Snout saw-like, flattened, with numerous lateral (rostral) teeth .... Family Pristidae, Sawfishes

Snout not saw-like, no rostral teeth 13

13. Two prominent dorsal fins 14

No dorsal fins or a single dorsal fin 19

14. Dorsal fins situated near the end of a slender tail, well behind pelvic fins .... Family Rajidae, Hardnose Skates
fig. 37; p. 152

Dorsal fins situated more anteriorly; tail not slender 15

15. Disc large relative to tail; dorsal fins close together .... Family Torpedinidae, Torpedo Rays
fig. 38; p. 102

Disc smaller relative to tail; dorsal fins well separated 16

16. Caudal fin with a well developed ventral lobe; pectoral and pelvic fins not touching 17

Caudal fin without a well defined lower lobe; pectoral and pelvic fins touching or overlapping 18

17. Head triangular; some small thorns present on back; two skin folds on spiracle .... Family Rhynchobatidae, Wedgefishes
fig. 39; p. 92

Head broadly rounded; back with strong ridges lined with large thorns; no folds on spiracle .... Family Rhinidae, Shark Rays
fig. 40; p. 90

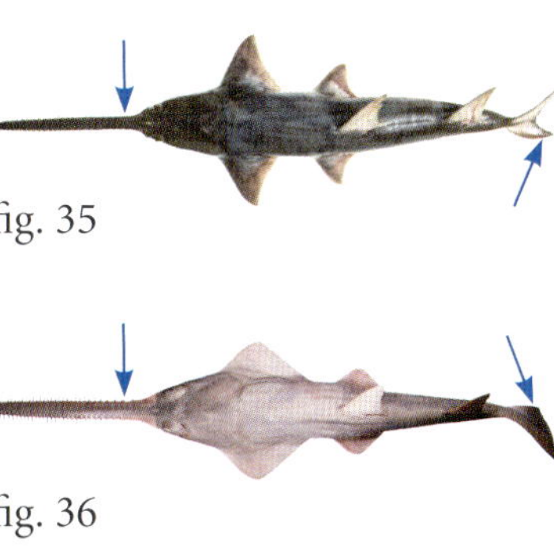

fig. 35

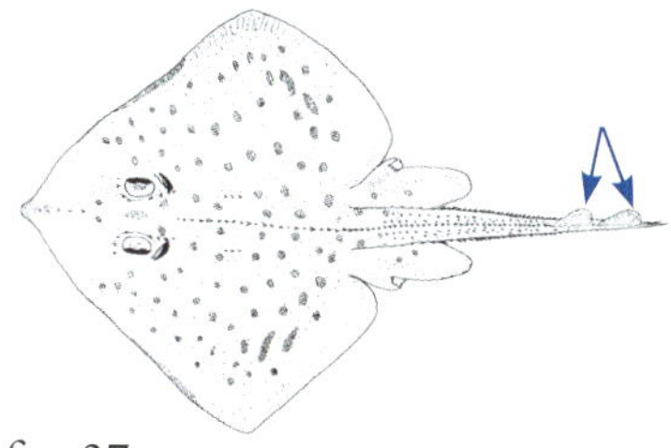

fig. 36

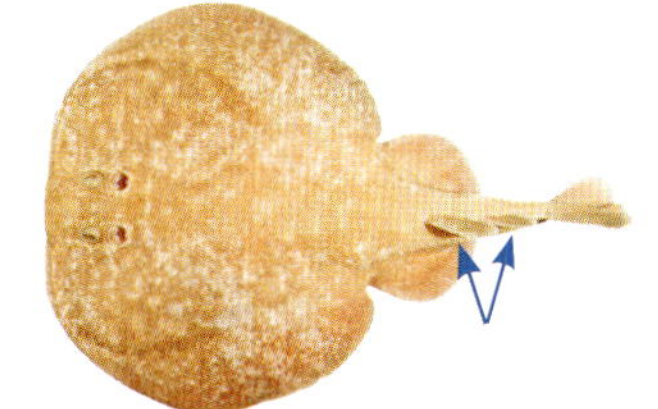

fig. 37

fig. 38

fig. 39

fig. 40

18. Snout triangular; body surface rough with thorns or fine denticles; no electric organs present on ventral surface .... Family Rhinobatidae, Shovelnose Rays

Snout broadly rounded; body surface entirely smooth; electric organs present (visible on ventral surface) .... Family Narcinidae*, Numbfishes fig. 45; p. 151

19. Anterior part of head extended beyond disc; eyes located laterally on side of head 20

Anterior part of head not extended beyond disc; eyes located dorsally and well inward from disc margin 22

20. A pair of long, paddle-like flaps (cephalic lobes) extending forward from sides of head; teeth minute, in many rows .... Family Mobulidae, Devilrays

No cephalic lobes, instead with a single, fleshy rostral lobe or a pair of broadly rounded lobes; teeth large, fewer than 10 rows in each jaw 21

21. Margin of rostral lobe rounded, not bilobed .... Family Myliobatidae, Eagle Rays

Margin of rostral lobe with a deep central notch

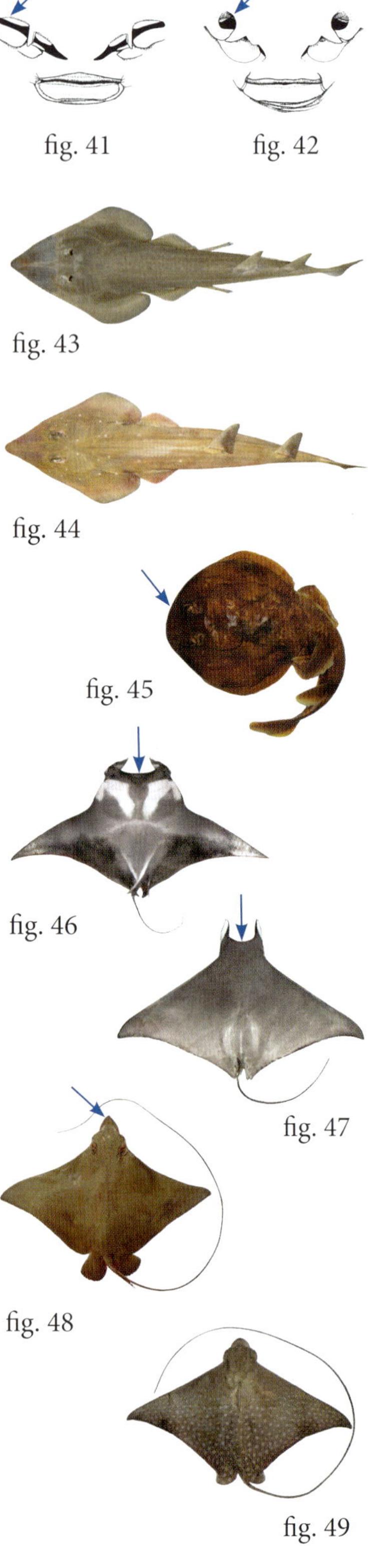

fig. 41   fig. 42

fig. 43

fig. 44

fig. 45

fig. 46

fig. 47

fig. 48

fig. 49

forming two rounded lobes ....
.... Family Rhinopteridae, Cownose Rays
fig. 50; p. 130

22. Disc very broad (more than 1.5 times wider than long); tail very short and filamentous .... Family Gymnuridae, Butterfly Rays
fig. 51; p. 120

Disc narrower (width less than 1.5 times length); tail moderately to very long with a thick base ... Family Dasyatidae, Stingrays

a. No skin folds on tail; base of tail narrow and typically rounded to slightly compressed in cross-section (fig. 52) .... b

Skin folds present on undersurface of tail (also sometimes on dorsal surface behind sting); base of tail relatively broad, distinctly depressed (fig. 53) c

b. No stinging spine on tail; numerous long, sharp thorns over entire disc .... Urogymnus* (fig. 54; p. 158)

One or more stinging spines on tail (scar visible if missing); no long, sharp thorns present on disc (low thorns near midline of disc in some species) .... Himantura (fig. 55; p. 106)

c. Disc oval or subcircular in shape; ventral skin fold on tail tall and extending to tail tip .... d

Disc roughly quadrangular in shape; ventral skin fold on tail lower and terminating well before tail tip (when undamaged) .... e

d. Disc oval in shape; dorsal surface with numerous blue spots over a brownish background; tail with a blue stripe along each side .... Taeniura* (fig. 56; p. 157)

Disc subcircular in shape; dorsal surface with black and white mottling; no blue spots or stripes on body or tail .... Taeniurops (fig. 57; p. 118)

e. Distance from cloaca to stinging spine exceeding half of disc width; ventral skin fold on tail relatively tall (maximum height equal to

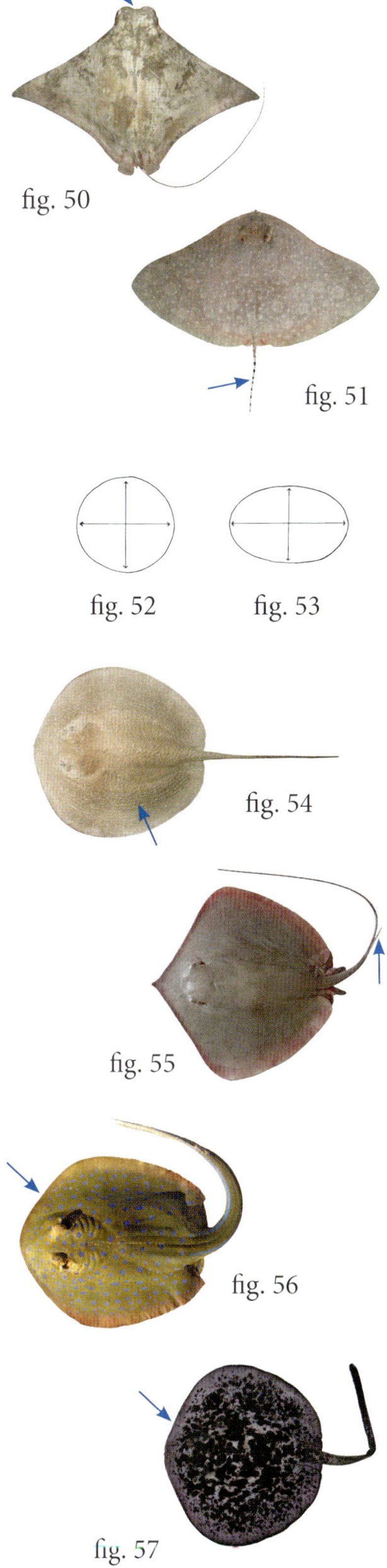

fig. 50

fig. 51

fig. 52   fig. 53

fig. 54

fig. 55

fig. 56

fig. 57

or exceeding spiracle length) ....
...................................*Pastinachus* (fig. 58; p. 114)

Distance from cloaca to stinging spine less than half of disc width; ventral skin fold relatively low (maximum height much less than spiracle length) .... f

f. Tail banded black and white behind sting; blue-spotted with dark mask-like band across eyes .... *Neotrygon** (fig. 59; p. 156)

Tail not banded beyond sting; not blue-spotted and without dark transverse band through eyes .... *Dasyatis* (fig. 60; p. 104)

fig. 58

fig. 59

fig. 60

Species information

Arabian Carpetshark

Chiloscyllium arabicum Goubanov *in* Goubanov & Shleib, 1980

Ventral head

SIZE	
	Maximum: 80 cm TL
	Maturity: males at 55–62 cm TL; females unknown
	At hatching: unknown; smallest free-swimming shark 10 cm TL

SIZE

Maximum: 80 cm TL

Maturity: males at 55–62 cm TL; females unknown

At hatching: unknown; smallest free-swimming shark 10 cm TL

KEY FEATURES

❶ Nasal barbels long

❷ Very elongate and thick precaudal tail

❸ Dorsal fins well separated, not falcate

❹ Prominent ridges on back before first dorsal fin and between dorsal fins

❺ Long, low anal fin just anterior to caudal fin

❻ Uniformly brown above, whitish below (juveniles and adults)

Worldwide distribution: From the Gulf eastwards to the southern tip of India.

Gulf occurrence: Widespread and common.

Habitat & biology: A coastal benthic species found on a variety of bottom types, including in and around coral reefs. Feeds on a range of prey including invertebrates (shrimps, crabs, squid and worms) and fish. The most important fish species consumed by this species in Kuwait is a mud-burrowing goby, although active pelagic fish (e.g. anchovies and mullet) were also taken. Oviparous; in aquaria, it has been recorded as having an annual reproductive cycle, producing four egg capsules at a time that are deposited on seabed objects such as hydroids and seaweed.

Conservation status: IUCN Red List: Near Threatened (assessed 2009).

Remarks: Often caught as trawl and gillnet bycatch, although highly resilient and able to survive long periods out of water. A specimen of *Chiloscyllium griseum* was recently reported from the shark landings in the United Arab Emirates, but since this record could not be verified, it has not been included in this guide.

References: Euzen (1987); Randall (1995); Jabado *et al.* (2014)

Underwater image from Kuwait

Carpet sharks are considered as especially low value in fish markets in the Gulf, and are often discarded at sea (Sharq market, Kuwait, April 2008)

Image details: Lateral: Bahrain (43 cm TL, 20th Aug 1985); Ventral head: Kuwait (adult male ~55 cm TL, 12th Apr 2011).

Tawny Nurse Shark

Nebrius ferrugineus (Lesson, 1831)

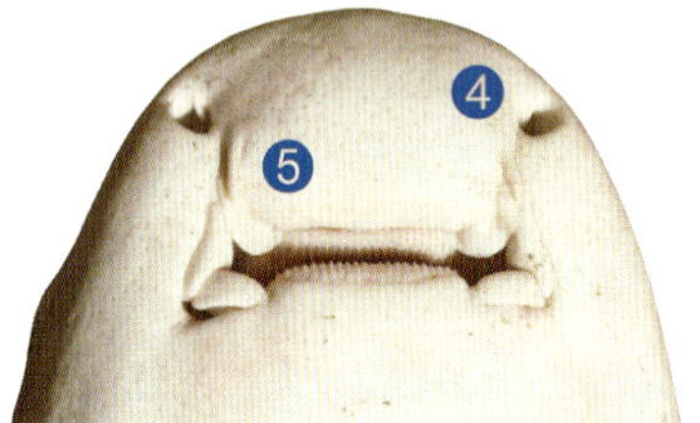

Ventral head

Worldwide distribution: Indo–West & Central Pacific, from southeastern Africa through to Tahiti and southern Japan.

Gulf occurrence: Apparently rare; recently confirmed from the shark landings in the United Arab Emirates.

Habitat & biology: A coastal species usually found around sandy, seagrass and coral reef areas; most active at night and rests in caves and crevices during the day. Depth ranges from the intertidal zone to at least 70 m. Feeds on a variety of bottom dwelling prey including cephalopods (particularly octopus), crabs, reef fishes, sea urchins and occasionally sea snakes. Aplacentally viviparous, with oophagy; litter size variable; 2–4 pups per litter, based on data from pregnant females at the Okinawa aquarium (Japan); in Australia, litters of up to 32 pups have been recorded.

Conservation status: IUCN Red List: Vulnerable (assessed 2003).

Remarks: Known to spin when hooked, and spit when removed from the water, making them difficult to handle. Commonly kept in aquaria due to their hardiness. Reported to have undergone severe declines in some regions.

References: Jabado *et al.* (2014)

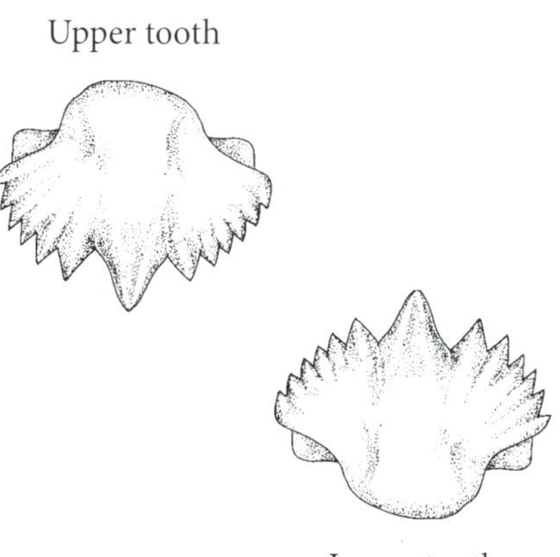

Image details: Lateral: Singapore (female 94 cm TL, 1800s); Ventral head: northern Australia (juvenile male 73 cm TL, 15ᵗʰ Aug 2000).

Ginglymostomatidae (Nurse Sharks)

Zebra Shark

Stegostoma fasciatum (Hermann, 1783)

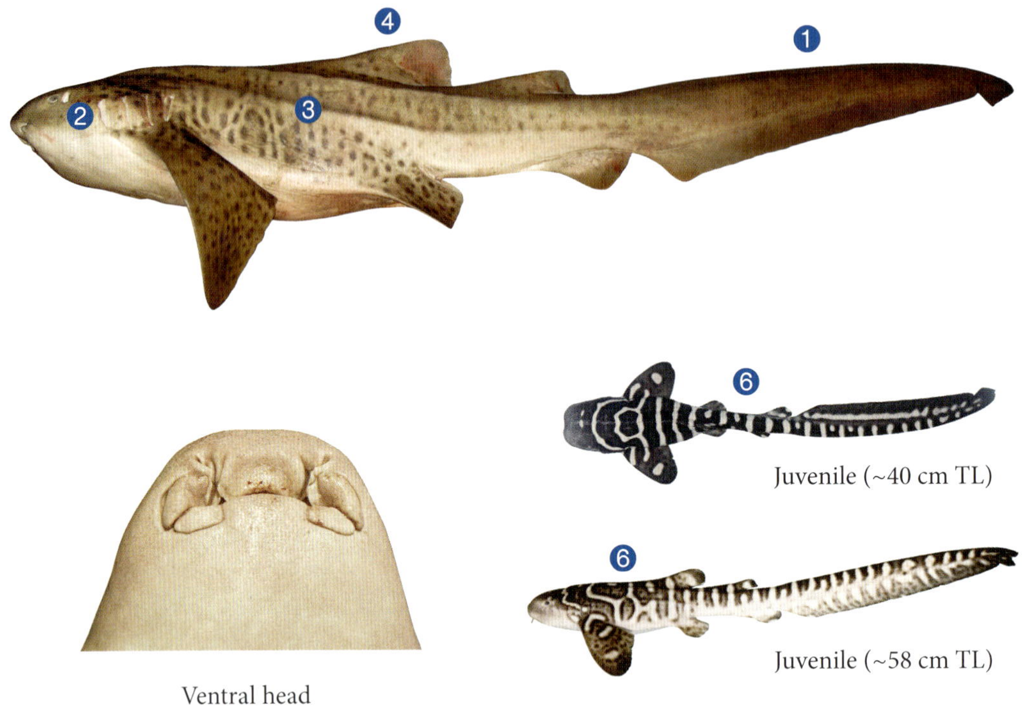

Juvenile (~40 cm TL)

Juvenile (~58 cm TL)

Ventral head

<table>
<tr><td rowspan="3">SIZE</td><td>Maximum: 354 cm TL</td></tr>
<tr><td>Maturity: males at 147–183 cm TL; females at 169–171 cm TL</td></tr>
<tr><td>At hatching: 20–36 cm TL</td></tr>
</table>

KEY FEATURES

❶ Caudal fin very long and blade-like

❷ Spiracle similar in size to eye

❸ Body with prominent ridges

❹ First dorsal fin much larger than second

❺ Adults yellowish with many dark brown spots

❻ Small juveniles dark brown with vertical white bars and spots

Worldwide distribution: Indo–West Pacific, from southern Africa through to New Caledonia and southern Japan.

Gulf occurrence: Patchy and rare. The only literature records are of single individuals from Kuwait and Bahrain in the 1970s/1980s.

Habitat & biology: Elsewhere in its range, this species is found in the intertidal zone to at least 62 m depth in habitats including coral reef and sand. A nocturnal species that hunts at night for benthic crustaceans, molluscs and small bony fishes; rests motionless on the bottom during the day. Mostly solitary, but also forms seasonal aggregations. Oviparous; a captive female in the Burj Al-Arab Aquarium (Dubai) was recently shown to produce pups in the absence of a male (parthenogenesis).

Conservation status: IUCN Red List: Vulnerable (assessed 2003).

Remarks: The common name of Zebra Shark relates to the colour pattern of small juveniles, which is possibly the result of mimicking banded sea snakes to avoid predation. Based on a genetic study in Australia, severe localized depletions of this species have been predicted.

References: Robinson *et al.* (2011); Dudgeon & White (2012)

Newborn Zebra Shark swimming near surface in shallow inshore, turbid waters off northwestern Australia

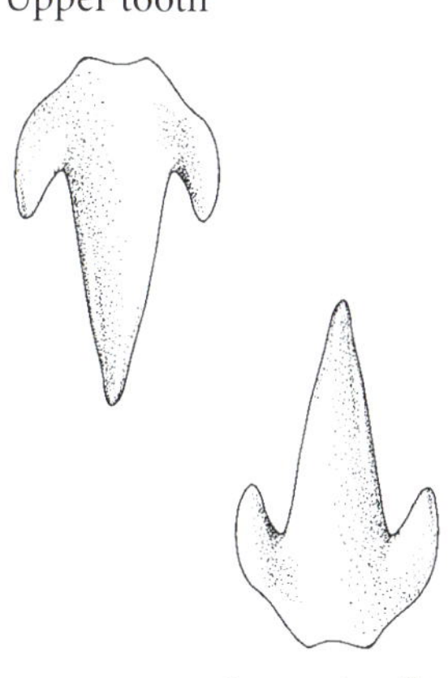

Upper tooth

Lower tooth

Image details: Lateral (adult) and ventral head: Dubai, United Arab Emirates (female ~120 cm TL, 18th Dec 2011); Juveniles: Bahrain (40.5 cm TL, 9th Feb 1977; 58 cm TL, 20th Jun 1984).

Whale Shark

Rhincodon typus Smith, 1828

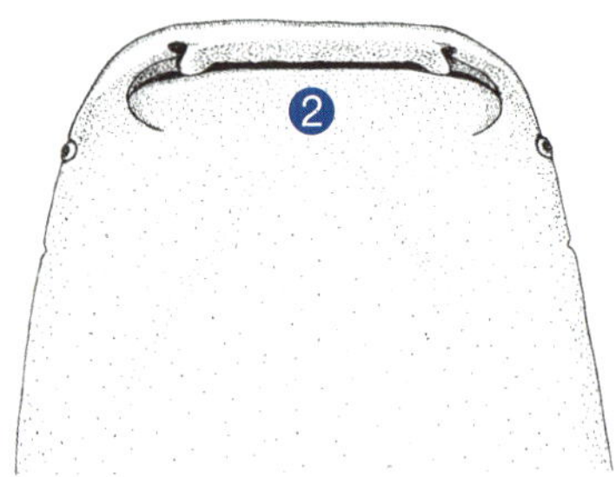

Ventral head

<table>
<tr><td rowspan="3">SIZE</td><td>Maximum: 17–21 m TL, but rarely above 12 m TL</td></tr>
<tr><td>Maturity: males at about 6 m TL and females at about 8 m TL</td></tr>
<tr><td>At birth: 55–64 cm TL, but may vary considerably</td></tr>
</table>

KEY FEATURES

❶ Head very broad and flattened

❷ Mouth very wide, almost terminal on head

❸ Adults huge, attaining more than 12 m TL

❹ Caudal peduncle depressed with a strong keel on each side

❺ Prominent ridges on dorsal surface and sides

❻ Body bluish grey with many yellowish spots, bars and stripes

Worldwide distribution: Worldwide in all tropical and warm temperate seas (except the Mediterranean).

Gulf occurrence: Widespread, with records from as far north as Iraq.

Habitat & biology: A highly migratory and pelagic species, which inhabits shallow coastal areas as well as the open ocean. Large aggregations of up to 100 animals have been recorded around gas platforms off Qatar. Feeds on planktonic crustaceans, fish eggs and small shoaling fishes. Aplacentally viviparous, in which egg cases are retained in utero until just prior to hatching; a pregnant female caught off Taiwan contained 304 near-term embryos.

Conservation status: IUCN Red List: Vulnerable (assessed 2005); listed on CITES Appendix II and on Appendix II of the Memorandum of Understanding of the Convention on Migratory Species (CMS).

Remarks: In the Gulf, juveniles sometimes become trapped in marinas, and have also been killed by ship strikes or entanglement in nets. In the 1930s, it was reported that Iranian fishermen hunted this species by jumping onto its back and inserting a large hook into its mouth.

References: Blegvad & Løppenthin (1944); Rowat & Brooks (2012); Robinson *et al.* (2013)

Whale Shark feeding off Qatar

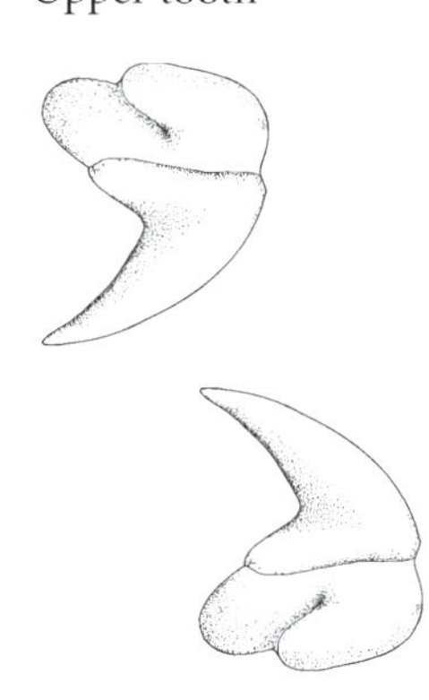

Upper tooth

Lower tooth

Image details: Lateral: Okinawa Churaumi Aquarium, Japan (adult male, 28th June 2013).

Sandtiger Shark

Carcharias taurus Rafinesque, 1810

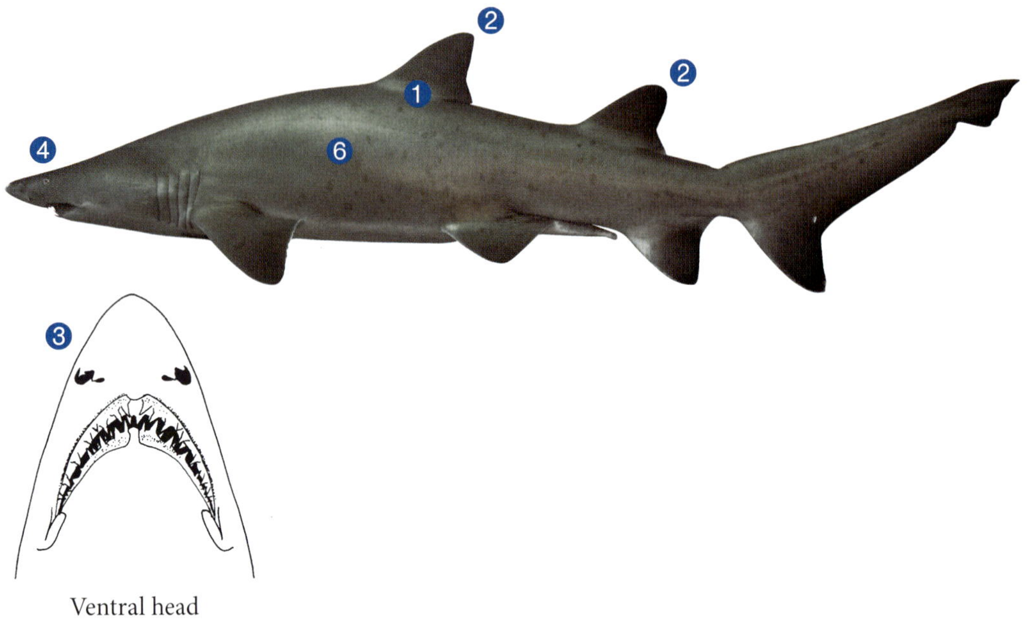

Ventral head

<table>
<tr><td rowspan="3">SIZE</td><td>Maximum: 430 cm TL</td></tr>
<tr><td>Maturity: both sexes at about 220 cm TL</td></tr>
<tr><td>At birth: 95–105 cm TL</td></tr>
</table>

KEY FEATURES

❶ First dorsal fin base closer to pelvic fins than pectoral fins

❷ Dorsal fins similar in size to anal fin

❸ Snout short and flattened

❹ Eyes very small, without moveable (nictitating) eyelid

❺ Teeth large with a long, narrow primary cusp and two lateral cusplets

❻ Body brownish, usually with scattered darker brown or reddish spots

Worldwide distribution: Atlantic and Indo-West Pacific Oceans.

Gulf occurrence: Patchy and rarely recorded, with single individuals from Kuwait, Qatar and Abu Dhabi Emirate (the latter caught from shallow waters of about 15 m depth).

Habitat & biology: A coastal species that occurs in a range of habitats including shallow bays and reefs. Can be solitary or occur in aggregations of up to 80 individuals. Feeds on bony fishes, crustaceans, squids, and other sharks and rays. Ovoviviparous, with intrauterine cannibalism, where the most developed embryo feeds on its siblings; only one embryo survives in each uterus at the end of the 9–12 month pregnancy.

Conservation status: IUCN Red List: Vulnerable (assessed 2009); Critically Endangered in the southwest Atlantic and eastern Australia.

Remarks: Aggregation sites are important for diving ecotourism in some areas (e.g. Australia). Sought after as an exhibit in public aquaria due to its large size, docile nature and long life span (over 30 years in aquaria). The pattern of spots on the flank has been used to identify individual sharks. Many populations have been reported to be seriously depleted.

References: Moore *et al.* (2007, 2012b); Jabado *et al.* (2013)

This dried specimen of *C. taurus*, in the stores of the Qatar National Museum, is one of the few records of this species in the Gulf (~200 cm TL, near Halul Island, Qatar, 1986)

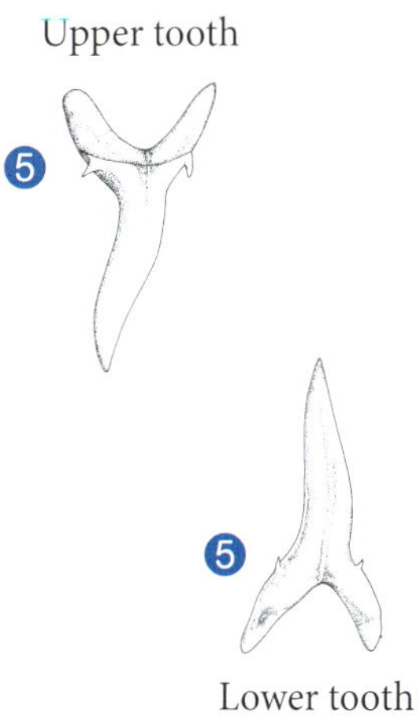

Image details: Lateral: eastern Australia (adult male, 28th Jul 2007).

Arabian Smoothhound

Mustelus mosis Hemprich & Ehrenberg, 1899

Ventral head

KEY FEATURES

❶ Teeth in both jaws broad and blunt (pavement-like)

❷ Upper labial furrows about equal in length to lowers

❸ Body uniformly greyish brown without white spots

❹ First dorsal fin white tipped

❺ Second dorsal and caudal fins black-tipped (more prominent in late-term embryos and newborns)

❻ Eyes dorsolateral with a strong ridge underneath

Worldwide distribution: Western Indian Ocean, from off KwaZulu-Natal, South Africa, the Red Sea and the Gulf to India and Sri Lanka.

Gulf occurrence: Widespread; frequently recorded in landings.

Habitat & biology: A bottom dwelling species that occurs in continental waters; elsewhere recorded at depths of 20 to 250 m. Feeds on crustaceans, molluscs and small bottom fishes. Viviparous, with yolk-sac placenta; producing 6–10 pups per litter.

Conservation status: IUCN Red List: Data Deficient (assessed 2009).

Remarks: Easily identifiable, as this is the only smoothhound (i.e. shark with flat pavement teeth) in the area. Does well in captivity. There is limited information on catches and biology of this species.

References: Moore *et al.* (2012a)

Triakidae (Houndsharks)

Litter of embryos from a 95 cm TL pregnant female,
Bahrain, April 2012

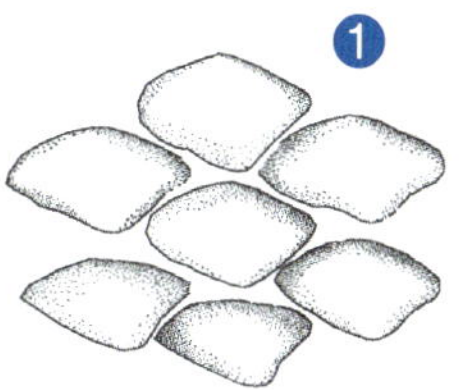

Upper teeth

Image details: Lateral and ventral head: Dubai, United Arab Emirates (adult male ~64 cm TL, 8th Oct 2012).

Hooktooth Shark

Chaenogaleus macrostoma (Bleeker, 1852)

Ventral head

<table>
<tr><td rowspan="3">SIZE</td><td>

Maximum: 100 cm TL (93 cm TL in the Gulf)

Maturity: females at 68–97 cm TL; males at 70–74 cm TL in
    the Gulf

At birth: at least 20 cm TL
</td></tr>
</table>

SIZE

Maximum: 100 cm TL (93 cm TL in the Gulf)

Maturity: females at 68–97 cm TL; males at 70–74 cm TL in the Gulf

At birth: at least 20 cm TL

KEY FEATURES

❶ Small spiracles present

❷ Gill slits large (more than twice length of eye)

❸ Snout obtusely wedge-shaped (in dorsoventral view)

❹ First dorsal fin not falcate

❺ Second dorsal fin usually with a prominent black tip

❻ Teeth protruding prominently from mouth when closed

❼ Anterior lower teeth with long, strongly hooked cusps

Worldwide distribution: Indo-West Pacific, from the Gulf of Aden through to Indonesia and Taiwan.

Gulf occurrence: Probably widespread; regularly occurs in fish markets, but in low numbers.

Habitat & biology: A coastal species found on the continental and insular shelves to at least 59 m depth. Feeds on cephalopods, crustaceans and small fishes. Biology poorly known; probably viviparous, with yolk-sac placenta; produces four pups per litter.

Conservation status: IUCN Red List: Vulnerable (assessed 2009).

Remarks: Weasel sharks, including this species, are likely to have been historically under-reported locally, due to their superficial resemblance to whaler sharks (Carcharhinidae).

References: Moore *et al.* (2012a)

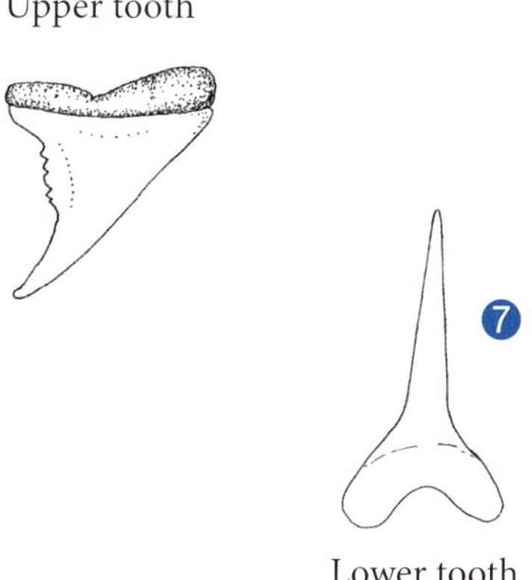

Image details: Lateral and ventral head: Bahrain (adolescent male ~70 cm TL, 7th Jun 2013).

Snaggletooth Shark

Hemipristis elongata (Klunzinger, 1871)

Ventral head

<table>
<tr><td rowspan="4">SIZE</td><td>Maximum: 240 cm TL</td></tr>
<tr><td>Maturity: males at 73–106 cm TL; females at 170–218 cm TL</td></tr>
<tr><td>At birth: 41–53 cm TL</td></tr>
</table>

<table>
<tr><td rowspan="7">KEY FEATURES</td><td>❶ Small spiracles present</td></tr>
<tr><td>❷ Gill slits large (more than twice length of eye)</td></tr>
<tr><td>❸ Snout bluntly rounded (in dorsoventral view)</td></tr>
<tr><td>❹ First dorsal fin strongly falcate</td></tr>
<tr><td>❺ Second dorsal fin without a prominent black tip</td></tr>
<tr><td>❻ Teeth protruding prominently from mouth when closed</td></tr>
<tr><td>❼ Anterior lower teeth with long, strongly hooked cusps</td></tr>
</table>

Worldwide distribution: Indo-West Pacific, from southeastern Africa to northern Australia and China.

Gulf occurrence: Patchy; only a few published records from Kuwait, Qatar and Bahrain.

Habitat & biology: A mainly coastal species found on continental and insular shelves, from the surface to at least 132 m depth. Feeds on cephalopods, rays, small sharks and bony fishes. Viviparous, with yolk-sac placenta; produces 2–11 (usually 6–8) pups per litter after a 7–8 month gestation period.

Conservation status: IUCN Red List: Vulnerable (assessed 2003); based on intensive fishing pressure throughout its range.

Remarks: Can be easily separated from other weasel sharks (*Chaenogaleus macrostoma* and *Paragaleus randalli*) by its strongly falcate fins and blunt snout.

References: Moore *et al.* (2010)

Upper tooth

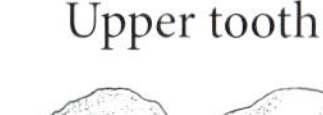

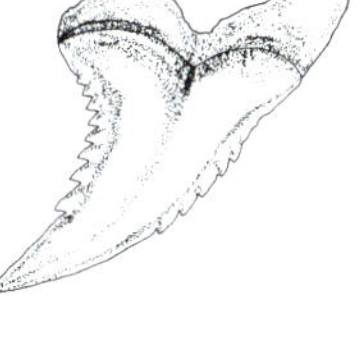

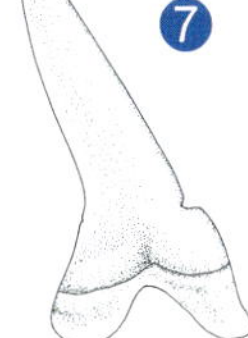

Lower tooth

Image details: Lateral and ventral head: Dubai, United Arab Emirates (juvenile male ~60 cm TL, 12th Oct 2012).

Slender Weasel Shark

Paragaleus randalli Compagno, Krupp & Carpenter, 1996

Ventral head

<table>
<tr><td rowspan="3" style="text-align:center">SIZE</td><td>Maximum: 84 cm TL</td></tr>
<tr><td>Maturity: males at 64–69 cm TL; females at about 71 cm TL</td></tr>
<tr><td>At birth: 29 cm TL</td></tr>
</table>

KEY FEATURES

❶ Small spiracles present

❷ Gill slits small (less than twice length of eye)

❸ Snout narrowly rounded (in dorsoventral view)

❹ Pectoral fins strongly falcate

❺ Second dorsal fin dark tipped, but without a prominent black apical blotch

❻ Teeth mostly concealed when mouth closed

❼ Anterior lower teeth with moderately long, erect cusps

❽ Two faint dusky stripes on ventral surface of snout

Worldwide distribution: Indo-West Pacific, from the Gulf through to Borneo and Taiwan.

Gulf occurrence: Probably widespread.

Habitat & biology: Found in shallow, coastal waters up to 18 m depth. Preliminary data suggest cephalopods form an important part of their diet. Viviparous, with yolk-sac placenta; producing two pups per litter (one in each uterus).

Conservation status: IUCN Red List: Near Threatened (assessed 2009).

Remarks: Previously thought to be found only in the northern Indian Ocean but specimens recently found off Borneo and off Taiwan. Regularly occurs in fish markets, although not in great numbers as with other weasel sharks. This species is likely to have been historically under-reported locally, due to superficial resemblance to whaler (Carcharhinidae) sharks.

References: White & Harris (2013)

Upper teeth

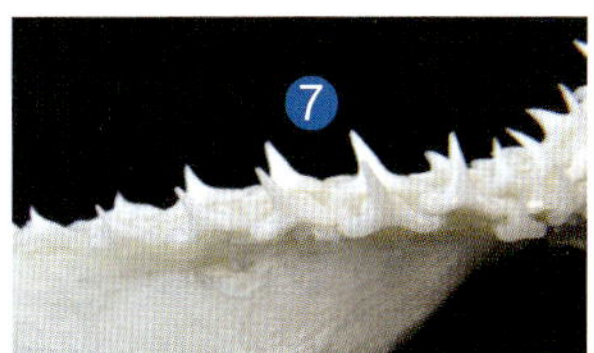

Stomachs of this species typically contain cephalopods (Abu Dhabi, 2012)

Lower teeth

Image details: Lateral and ventral head: Dubai, United Arab Emirates (adult male 64.6 cm TL, 12th Oct 2012); Upper and lower teeth: Taiwan (female 81 cm TL).

Graceful Shark

Carcharhinus amblyrhynchoides (Whitley, 1934)

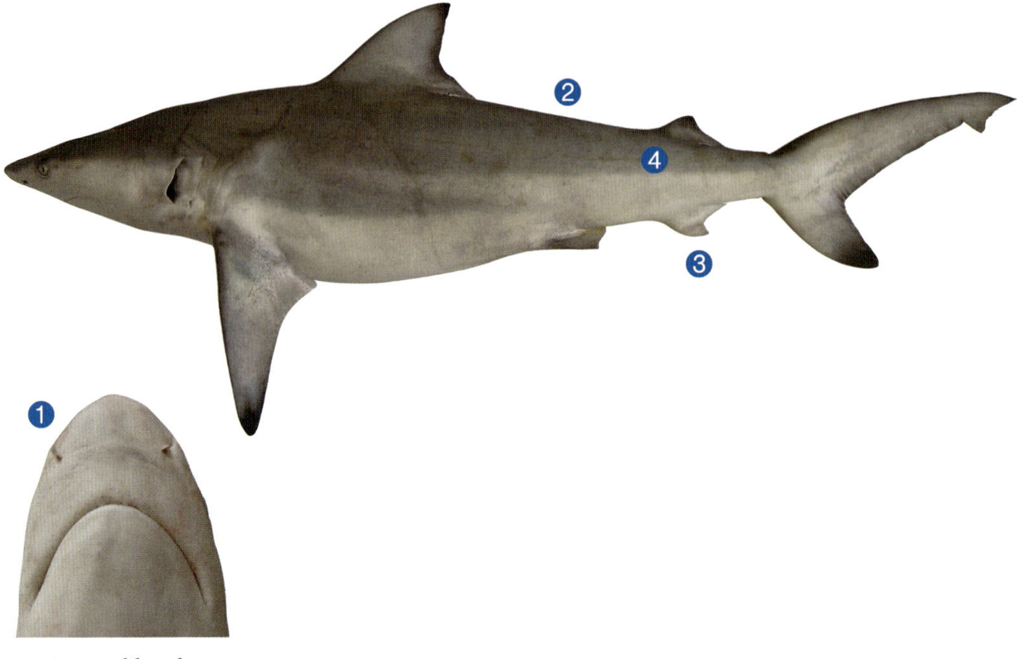

Ventral head

<table>
<tr><td rowspan="3">SIZE</td><td>Maximum: 178 cm TL</td></tr>
<tr><td>Maturity: both sexes at 104–115 cm TL</td></tr>
<tr><td>At birth: 50–60 cm TL</td></tr>
</table>

KEY FEATURES

❶ Snout moderately short and somewhat pointed

❷ No interdorsal ridge

❸ Anal fin plain, all other fins with black or dusky tips

❹ Second dorsal-fin origin level with anal-fin origin

❺ Upper and lower teeth with slender, erect, finely serrated cusps, without basal cusplets

❻ Less than 82 precaudal vertebrae

Worldwide distribution: Indo-West Pacific, from the Gulf of Aden to northern Australia and the Philippines.

Gulf occurrence: Likely to be widespread, but does not appear to be common.

Habitat & biology: A coastal species, inhabits coastal waters to a depth of at least 50 m. It feeds mainly on bony fishes and occasionally on crustaceans and cephalopods. Viviparous, with yolk-sac placenta; producing litters of 2–8 pups after a 9–10 month gestation (in northern Australian populations).

Conservation status: IUCN Red List: Near Threatened (assessed 2009).

Remarks: This widespread but poorly known species is commonly confused with (and genetically very similar to) the Smoothtooth Blacktip Shark *Carcharhinus leiodon* (p. 58) and the Common Blacktip Shark *Carcharhinus limbatus* (p. 62) but differs in fin colouration, vertebral counts, snout shape and dentition.

References: Moore *et al.* (2011)

Carcharhinidae (Whaler Sharks)

Comparison of *C. leiodon* (bottom; female, 85 cm TL) and *C. amblyrhynchoides* (top; female, 83 cm TL) showing difference in fin markings and coloration; Sharq market, Kuwait, April 2011.

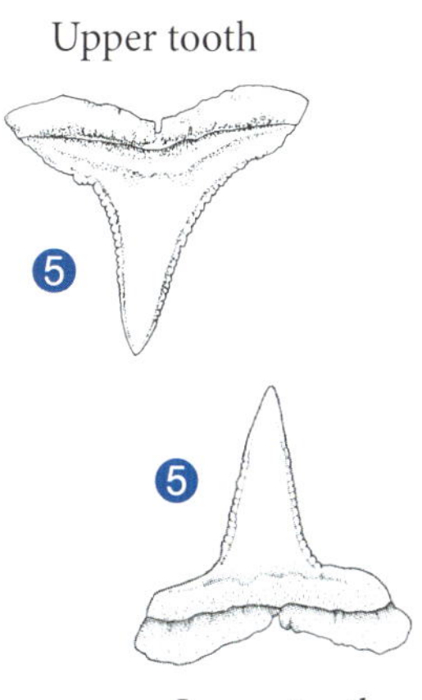

Image details: Lateral and ventral head: Kuwait (~90 cm TL, 22nd Apr 2008).

Grey Reef Shark

Carcharhinus amblyrhynchos (Bleeker, 1856)

Ventral head

<table>
<tr><td rowspan="1">SIZE</td><td>

Maximum: 255 cm TL, but rarely above 190 cm TL

Maturity: males at 130–145 cm TL; females at 120–142 cm TL

At birth: 45–60 cm TL

</td></tr>
</table>

KEY FEATURES

❶ Snout moderately rounded

❷ Usually no interdorsal ridge (rarely a weak ridge present)

❸ Caudal fin with a broad black posterior margin

❹ Second dorsal-fin origin level with anal-fin origin

❺ First dorsal-fin origin level with pectoral-fin free rear tips

❻ Upper teeth triangular, slightly oblique, serrated, with notch on one edge

❼ Lower teeth narrow, more erect, weakly serrate

Worldwide distribution: Indo-Pacific, from southeastern Africa through to the Galápagos Islands.

Gulf occurrence: Apparently very rare, with the only known published record being a single underwater photograph record from Jana Island (Saudi Arabia) in the 1970s. Further records are of interest.

Habitat & biology: An inshore species which has a strong association with coral reefs, particularly near dropoffs; occasionally in lagoons and open oceans; from the surface to at least 60 m depth, but has been reported to depths of 1,000 m. Feeds on bony fishes, cephalopods and crustaceans. Viviparous, with yolk-sac placenta; producing 1–6 pups per litter. A gestation period of 9–12 months was recorded in New Caledonian waters.

Conservation status: IUCN Red List: Near Threatened (assessed 2009).

Remarks: *Carcharhinus wheeleri* Garrick 1982, described from the Red Sea off Egypt, has been considered a synonym of this species. Some authors have recently considered it a valid species and taxonomic revision is required to determine the validity of this species. A white-tipped first dorsal fin is the main character which has been used to separate *C. wheeleri* from *C. amblyrhynchos*, but this feature is often also observed in the latter species throughout its range and thus is not reliable.

References: Fourmanoir (1976); Moore *et al.* (2010); Naylor *et al.* (2012)

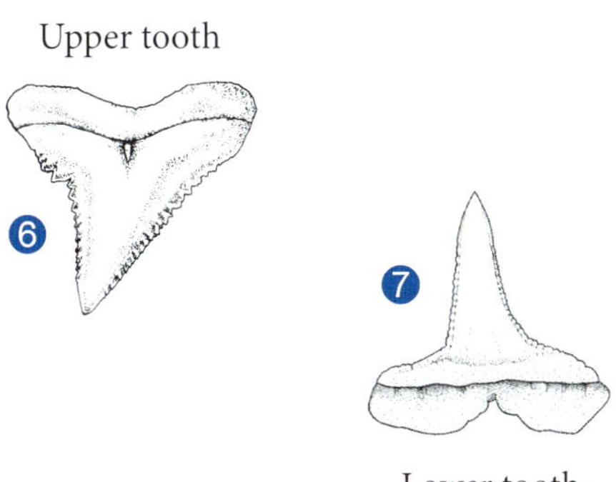

Image details: Lateral: Jeddah, Saudi Arabia (76 cm TL, 10th May 1982); Ventral head: Indonesia (female 70 cm TL, 13th Oct 2004).

Pigeye Shark

Carcharhinus amboinensis (Müller & Henle, 1839)

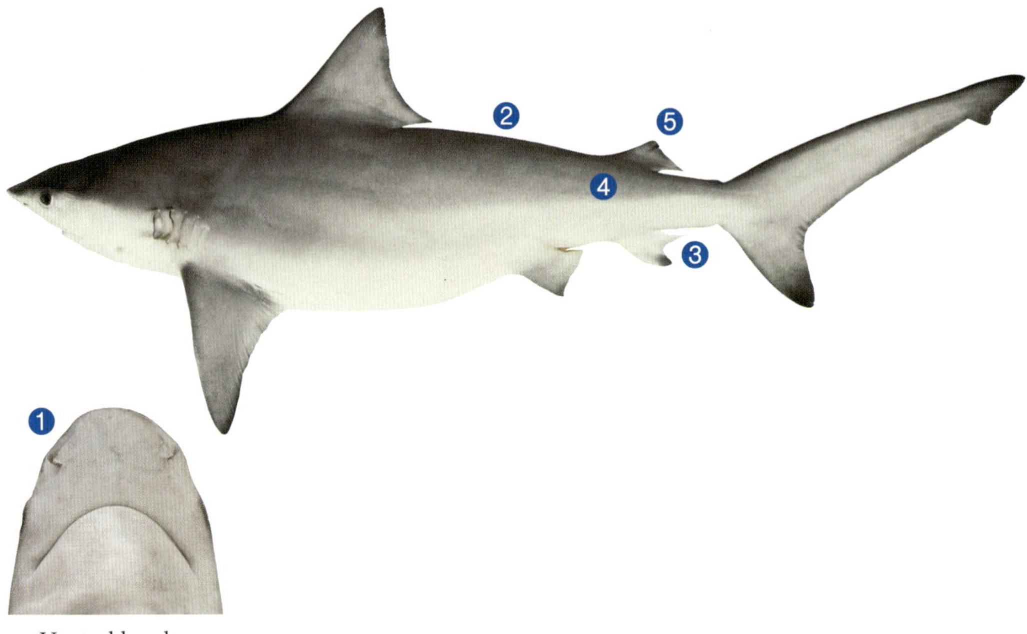

Ventral head

SIZE

Maximum: 280 cm TL

Maturity: males at 195–210 cm TL (206–227 cm TL in the Gulf); females mature at 198–223 cm TL

At birth: 60–72 cm TL (as small as 57 cm TL in the Gulf)

KEY FEATURES

❶ Snout short and blunt

❷ No interdorsal ridge

❸ Notch on posterior margin of anal fin acute (less than a right angle)

❹ Second dorsal-fin origin level with anal-fin origin

❺ Second dorsal fin small, less than a third height of first

❻ Fins without markings (juveniles may have dusky fin tips)

❼ Upper and lower teeth broadly triangular and serrated

Worldwide distribution: Indo-West Pacific, from southeastern Africa to northern Australia and Papua New Guinea; a record from Nigeria is doubtful.

Gulf occurrence: Probably widespread.

Habitat & biology: A coastal species inhabiting shallow waters, including bays and estuaries. Feeds on bony fishes, crustaceans, molluscs, and small sharks and rays. Viviparous, with yolk-sac placenta; producing 6–13 pups per litter after a 12 month gestation period.

Conservation status: IUCN Red List: Data Deficient (assessed 2009).

Remarks: Molecular analyses suggest this is possibly a species complex. Similar to the Bull Shark *Carcharhinus leucas* (p. 60) but the relative heights of the dorsal fins can be used as a reliable field character to separate these two species. Other differences include the angle of the anal-fin posterior margin, number of precaudal vertebrae (89–95 vs. 101–123 in *C. leucas*), and number of lower jaw teeth (11 vs. 12 teeth on each side).

References: Naylor *et al.* (2012); Moore *et al.* (2012a)

A large Pigeye Shark landed at Mina Zayed Port, Abu Dhabi, April 2010

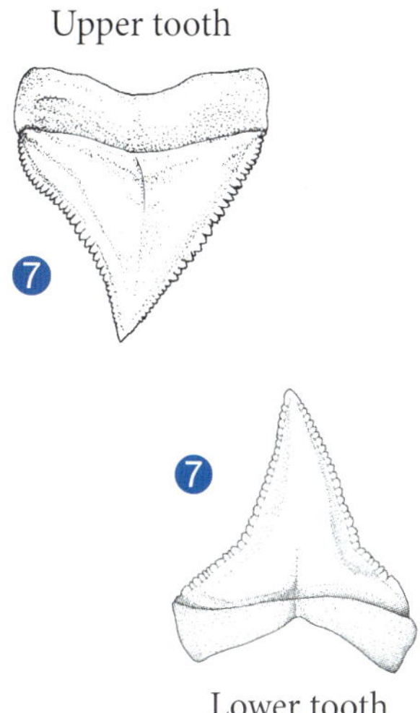

Image details: Lateral: Bahrain (81 cm TL, 1st Nov 1983); Ventral head: Dubai, United Arab Emirates (juvenile ~70 cm TL, 7th Oct 2012).

Spinner Shark

Carcharhinus brevipinna (Müller & Henle, 1839)

Ventral head

Juvenile

<table>
<tr><td rowspan="3">SIZE</td><td>Maximum: 300 cm TL</td></tr>
<tr><td>Maturity: males at 190–200 cm TL; females at 198–223 cm TL</td></tr>
<tr><td>At birth: 60–80 cm TL</td></tr>
</table>

KEY FEATURES

❶ Snout long and pointed

❷ No interdorsal ridge

❸ Fins plain in small juveniles; specimens over about 1 m TL with black tips (more distinct in larger individuals)

❹ Second dorsal-fin origin level with anal-fin origin

❺ First dorsal-fin origin slightly behind pectoral-fin free rear tips

❻ Upper labial furrows conspicuous

❼ Upper and lower teeth with low, narrow, erect cusps

Worldwide distribution: Warm temperate and tropical waters of the Atlantic, Indian and West Pacific Oceans.

Gulf occurrence: Probably widespread.

Habitat & biology: Occurs from the surface to at least 100 m depth. Feeds on a variety of small bony fishes and cephalopods. Viviparous, with yolk-sac placenta; producing 3–15 pups per litter. A two-year reproductive cycle, with a gestation period of 13–18 months was reported from KwaZulu-Natal in South Africa.

Conservation status: IUCN Red List: Near Threatened (assessed 2009).

Remarks: A schooling, migratory species which has often leaps spinning out of the water after making feeding runs through baitfish schools. Age at maturity is 8–10 years based on studies on populations in South Africa and Taiwan. Similar to the Common Blacktip Shark *Carcharhinus limbatus* (p. 62) but has a lower first dorsal fin, cusps on teeth lower, longer labial furrows, more slender body, and different fin markings.

References: None

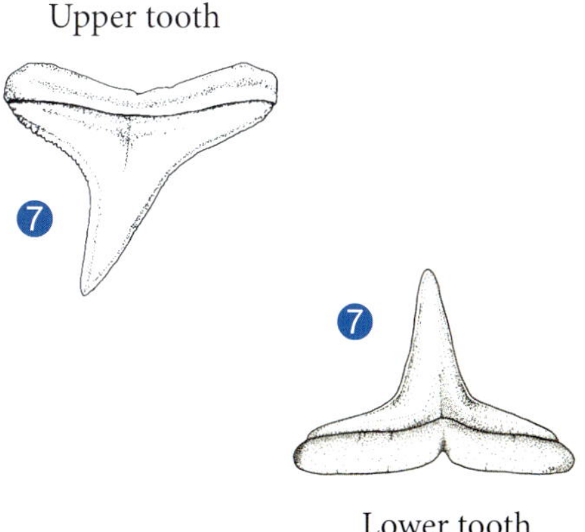

Image details: Lateral: Dubai, United Arab Emirates (female ~240 cm TL, 7th Oct 2012); Ventral head: Borneo, Malaysia (juvenile male 118 cm TL, 28th Jun 2002); Juvenile: Borneo, Malaysia (juvenile male 91 cm TL, 6th May 2002).

Whitecheek Shark

Carcharhinus dussumieri (Müller & Henle, 1839)

Ventral head

Maximum: 100 cm TL

Maturity: In the Gulf, 50% of males mature by 72 cm TL and females by 79 cm TL (69 and 68 cm TL off the Iranian coast of the Gulf of Oman)

At birth: 36–38 cm TL in the Gulf

❶ Second dorsal fin with a large black blotch, other fins plain

❷ First dorsal fin broadly triangular, not falcate

❸ Snout moderately long and narrowly parabolic

❹ Second dorsal-fin origin level with anal-fin origin

❺ Upper teeth oblique, deeply notched, with serrated basal cusplets

❻ Lower teeth narrower, slightly oblique, without cusplets

Worldwide distribution: From the Gulf eastwards to India.

Gulf occurrence: Widespread and often highly abundant.

Habitat & biology: A coastal species found in shallow water to a depth of at least 170 m. Feeds on small fish, crustaceans and squids. Viviparous, with yolk-sac placenta; producing 1–5 (usually 2) pups per litter; pupping in Iranian waters of the Sea of Oman has been reported to peak in June to July.

Conservation status: IUCN Red List: Near Threatened (assessed 2003).

Remarks: Previously thought to be a wide-ranging species in the Indo-West Pacific, but a recent taxonomic revision revealed it is restricted to the northwest Indian Ocean. Very similar to the newly described, but much rarer, Human's Whaler Shark *Carcharhinus humani* (p. 56) but differs in having a more flattened and pointed snout (lateral view), a broad and not falcate first dorsal fin, and coarsely serrated basal cusplets on upper teeth (vs. smooth basal cusplets). Mercury concentrations in this shark have been reported as being potentially harmful to humans. Tapeworm parasites in this species are thought to act as a 'sink' or buffer to toxic heavy metals.

References: Assadi (2001); Malek *et al.* (2007); Delshad *et al.* (2012); Moore *et al.* (2012a); White (2012)

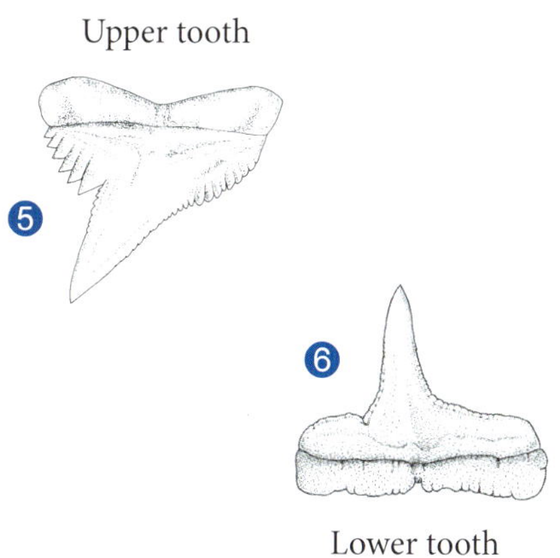

Image details: Lateral and ventral head: Kuwait (adult male ~82 cm TL, 12th Apr 2011).

Silky Shark

Carcharhinus falciformis (Müller & Henle, 1839)

Ventral head

<table>
<tr><td rowspan="1">SIZE</td><td>

Maximum: 330 cm TL

Maturity: males mature at 180–217 cm TL and females at 180–230 cm TL

At birth: 56–87 cm TL

</td></tr>
<tr><td>KEY FEATURES</td><td>

❶ First dorsal-fin relatively low, origin well behind pectoral-fin rear tips

❷ Interdorsal ridge present

❸ Second dorsal-fin low, its inner margin very long (1.6–3.0 times its height)

❹ Snout relatively long, narrowly rounded (in ventral view)

❺ Upper teeth narrow with one edge prominently notched

❻ Lower teeth narrow, upright

</td></tr>
</table>

Worldwide distribution: Circumglobal in all tropical seas.

Gulf occurrence: Apparently rare; recently confirmed from off Dubai, United Arab Emirates.

Habitat & biology: An epipelagic and oceanic species, preferring deeper waters close to landmasses, but also found well offshore and occasionally in coastal waters; from the surface to at least 500 m depth. Feeds on cephalopods, crabs and bony fishes such as tuna, mackerel and sardines. Viviparous, with a yolk-sac placenta; 1–16 (usually >6) pups per litter; no apparent reproductive seasonality.

Conservation status: IUCN Red List: Near Threatened (assessed 2009).

Remarks: Common in Omani fisheries and sometimes present in Gulf markets, such as Dubai, after being transported overland from Oman. One of the most frequently caught species in commercial longline fisheries in many tropical areas.

References: Henderson *et al.* (2009); Hall *et al.* (2012); Jabado *et al.* (2014)

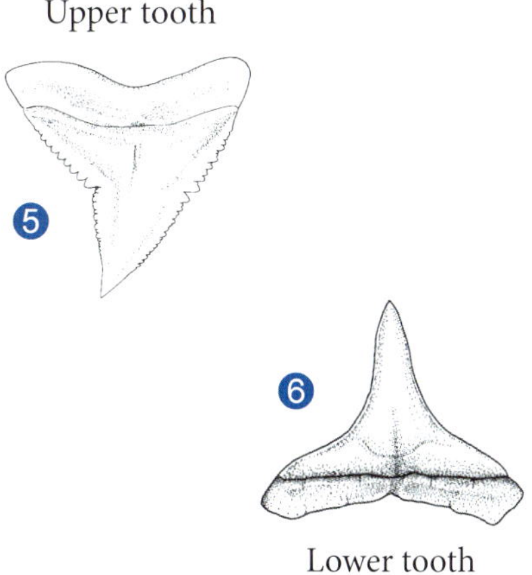

Image details: Lateral: Dubai, United Arab Emirates (female ~120 cm TL, 26th Apr 2013); Ventral head: Indonesia (juvenile male 115 cm TL, 29th Mar 2006).

Human's Whaler Shark

Carcharhinus humani White & Weigmann, 2014

Ventral head

<table>
<tr><td rowspan="3">SIZE</td><td>Maximum: 92 cm TL</td></tr>
<tr><td>Maturity: males at 75–83 cm TL; females above 75 cm TL</td></tr>
<tr><td>At birth: 35–45 cm TL (based on South African specimens)</td></tr>
</table>

KEY FEATURES

❶ Second dorsal fin with a large black blotch, other fins plain

❷ First dorsal fin narrowly triangular and falcate

❸ Snout moderately long and narrowly parabolic

❹ Second dorsal-fin origin level with anal-fin origin

❺ No interdorsal ridge

❻ Upper teeth oblique, deeply notched, with smooth basal cusplets

❼ Lower teeth narrower, slightly oblique, without cusplets

Worldwide distribution: Western Indian Ocean, from the Natal coast of South Africa to the Gulf, also Socotra Islands and Seychelles.

Gulf occurrence: Poorly known; only known from two confirmed records from Bahrain and Kuwait; probably more widespread.

Habitat & biology: Occurs in depths of less than 40 m, although one specimen off Madagascar was caught at the surface in water over 1,200 m depth. Feeds mostly on bony fishes, cephalopods and crustaceans. Viviparous, with yolk-sac placenta; 6 pregnant females off South Africa contained either 1 or 2 embryos.

Conservation status: IUCN Red List: Not evaluated.

Remarks: Recently described as a new species; previously misidentified as *Carcharhinus sealei* (Pietschmann, 1913), but differs in having a taller second dorsal fin and a more strongly demarcated black marking on second dorsal fin. Very similar to the Whitecheek Shark *Carcharhinus dussumieri* (p. 52) but snout tip more rounded and less depressed in lateral view, first dorsal fin falcate, more vertebrae, and basal cusplets on upper teeth without serrations.

References: White & Weigmann (2014)

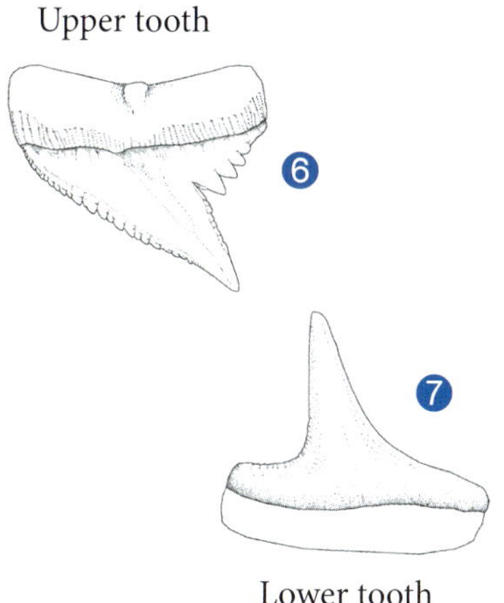

Image details: Lateral: Bahrain (adult male 72 cm TL, 9th Feb 1977); Ventral head: Socotra Islands (adult male 83 cm TL, 15th Jan 1989).

Smoothtooth Blacktip Shark

Carcharhinus leiodon Garrick, 1985

Ventral head

<table>
<tr><td rowspan="3">SIZE</td><td>Maximum: 165 cm TL</td></tr>
<tr><td>Maturity: males between 90–120 cm TL; females by at least 131 cm TL</td></tr>
<tr><td>At birth: 35–51 cm TL</td></tr>
</table>

SIZE

Maximum: 165 cm TL

Maturity: males between 90–120 cm TL; females by at least 131 cm TL

At birth: 35–51 cm TL

KEY FEATURES

❶ Snout short and bluntly pointed

❷ No interdorsal ridge

❸ All fins, including anal fin, with black tips

❹ Second dorsal-fin origin level with anal-fin origin

❺ A broad black margin on anterior margin of caudal fin

❻ Upper and lower teeth with slender, erect, smooth-edged cusps (fine serrations on adults)

❼ Dorsal surface often with a greenish tinge

Worldwide distribution: Restricted to the Gulf and the Arabian Sea (off Yemen and southern Oman).

Gulf occurrence: Known from off Kuwait and possibly the United Arab Emirates.

Habitat & biology: A coastal species with uncertain habitat requirements, but possibly associated with estuarine environments. Feeds on bottom-associated fish, including catfish. Viviparous, with yolk-sac placenta; the only two litters recorded contained 4 and 6 embryos.

Conservation status: IUCN Red List: Vulnerable (assessed 2005).

Remarks: One of the rarest whaler sharks in the world. Until its rediscovery in the fish markets of Kuwait in 2008, this species was known to science from only the holotype collected in Yemen in 1902. Very similar to the Graceful Shark *Carcharhinus amblyrhynchoides* (p. 44) but differs in having a distinctly black-tipped anal fin, a prominent black caudal-fin anterior margin, and teeth with smooth cusps.

References: Moore *et al.* (2011); Moore *et al.* (2013)

For over 100 years, this single preserved individual in the Naturhistorisches Museum in Vienna was the only Smoothtooth Blacktip Shark known to science, until its rediscovery in Kuwait in 2008

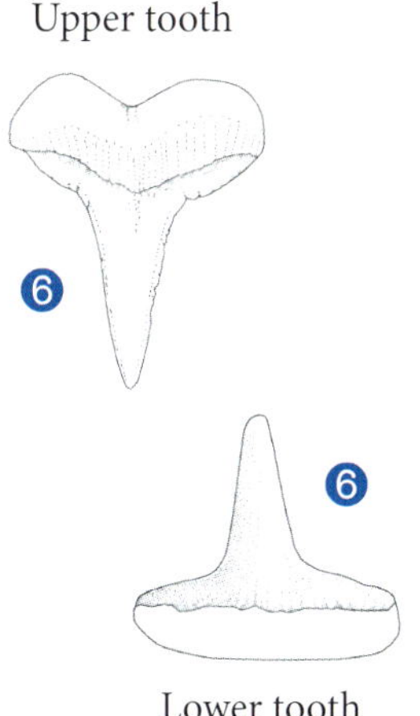

Upper tooth

Lower tooth

Image details: Lateral and ventral head: Kuwait (juvenile male 67 cm TL, 22nd Apr 2008).

Bull Shark

Carcharhinus leucas (Müller & Henle, 1839)

Ventral head

<table>
<tr><td rowspan="1">SIZE</td><td>

Maximum: 340 cm TL

Maturity: males at 157–226 cm TL; females at 180–230 cm TL

At birth: 55–81 cm TL

</td></tr>
</table>

KEY FEATURES

❶ Snout short and blunt

❷ No interdorsal ridge

❸ Notch on posterior margin of anal fin not acute (more than a right angle)

❹ Second dorsal-fin origin level with anal-fin origin

❺ Second dorsal fin relatively larger, more than a third height of first dorsal fin

❻ Fins without markings (juveniles may have dusky fin tips)

❼ Upper broadly triangular and strongly serrate

❽ Lower teeth narrowly triangular, erect

Worldwide distribution: Cosmopolitan in most warm temperate and tropical waters, including freshwater rivers and lakes.

Gulf occurrence: Adults and larger juveniles are likely to be widespread; young may be limited to estuarine-influenced areas, such as northern Kuwait, Iraq and the westernmost parts of Iran.

Habitat & biology: A coastal species found in marine, brackish and fresh waters ranging in depths from less than a metre to at least 152 m. Able to tolerate very low salinity environments and it has been recorded as far upstream as Baghdad in Iraq. Feeds on a diverse range of prey including bony fish, crustaceans, turtles, birds, marine mammals, small sharks and stingrays. Viviparous, with yolk-sac placenta; producing 1–13 pups per litter. A 10–11 month gestation period has been reported for this species in the Eastern Atlantic Ocean. The threatened Tigris-Euphrates-Karun system may represent a regionally important nursery area for this species.

Conservation status: IUCN Red List: Near Threatened (assessed 2009).

Remarks: The presence of sharks in the Tigris-Euphrates-Karun river system has been known for hundreds of years, and this species is likely to have been responsible for a number of human injuries and fatalities in Iraq and Iran. Similar to the Pigeye Shark *Carcharhinus amboinensis* (p. 48) but differs most clearly in relative heights of dorsal fins; other differences include the angle of the anal-fin posterior margin, number of precaudal vertebrae (101–123 vs. 89–95 in *C. amboinensis*), and number of lower jaw teeth (12 vs. 11 teeth on each side).

References: Coad & Papahn (1988); Coad & Al Hassan (1989); Cliff & Dudley (1991)

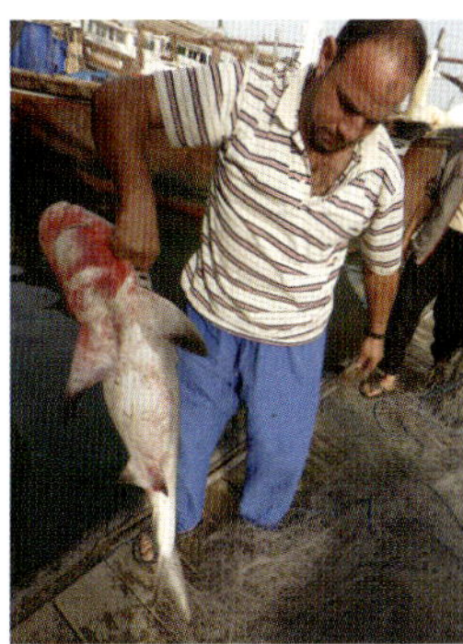

Bull Shark gillnet bycatch at Sharq fish
market, Kuwait, 5th Apr 2011

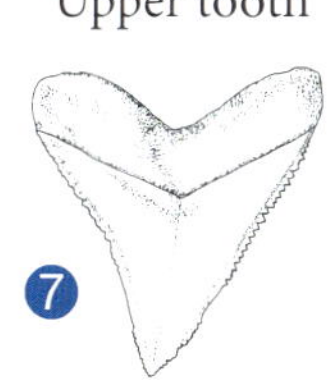

Upper tooth

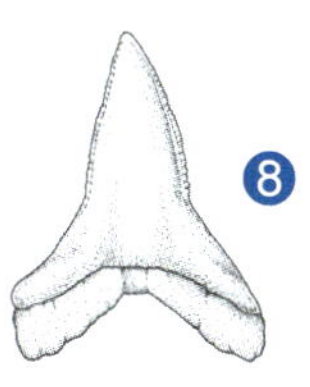

Lower tooth

Image details: Lateral and ventral head: Kuwait (~105 cm TL, 10th Apr 2008).

Common Blacktip Shark

Carcharhinus limbatus (Müller & Henle, 1839)

Ventral head

<table>
<tr><td rowspan="3">SIZE</td><td>Maximum: 255 cm TL</td></tr>
<tr><td>Maturity: males at 135–184 cm TL (164–184 cm TL in the Gulf) and females at 120–190 cm TL</td></tr>
<tr><td>At birth: 38–72 cm TL</td></tr>
</table>

KEY FEATURES

❶ Snout long and pointed

❷ No interdorsal ridge

❸ Fins black-tipped in small juveniles (plain or dusky in adults)

❹ Second dorsal-fin origin level with anal-fin origin

❺ First dorsal-fin origin about level with pectoral-fin insertion

❻ Upper labial furrows inconspicuous

❼ Upper and lower teeth with tall, narrow, erect cusps

Worldwide distribution: Cosmopolitan in all warm temperate and tropical waters.

Gulf occurrence: Widespread; probably the most common of the larger shark species within the Gulf.

Habitat & biology: Widespread in a range of habitats on continental and insular shelves. Feeds on bony fishes, crustaceans, cephalopods, as well as sharks and rays. Viviparous, with yolk-sac placenta; producing 1–10 pups per litter. A gestation period of 10–12 months was reported for South African populations.

Conservation status: IUCN Red List: Near Threatened (assessed 2009).

Remarks: A fast-swimming shark that can form large aggregations. Males and females attain maturity at 5–6 and 6–7 years, respectively. Similar to the Spinner Shark *Carcharhinus brevipinna* (p. 50) but has a taller first dorsal fin, higher cusps on teeth, more robust body, and different fin markings. Molecular analyses suggest that the Indo-Pacific and Atlantic populations may represent separate species.

References: Dudley & Cliff (1993); Naylor *et al.* (2012)

Underwater image of a juvenile Common Blacktip Shark off Kuwait, April 2008

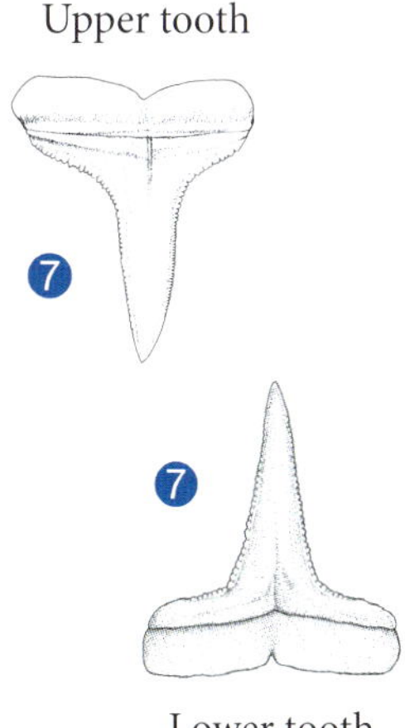

Image details: Lateral and ventral head: Dubai, United Arab Emirates (~110 cm TL, 25th May 2012).

Hardnose Shark

Carcharhinus macloti (Müller & Henle, 1839)

Ventral head

SIZE

Maximum: 110 cm TL

Maturity: males at 69–83 cm TL (75–83 cm TL in the Gulf);
females at 70–89 cm TL

At birth: 40–50 cm TL

KEY FEATURES

❶ Second dorsal-fin origin about level with anal-fin midbase

❷ Snout long and pointed with a noticeably hard rostrum

❸ No interdorsal ridge

❹ First dorsal-fin inner margin very long (about two thirds of fin base)

❺ Upper teeth narrow, oblique or erect, with large basal serrations

❻ Lower teeth more slender, erect and smooth

Worldwide distribution: Indo-West Pacific, from eastern Africa through to northern Australia and Taiwan.

Gulf occurrence: Apparently widespread, but patchy and not common.

Habitat & biology: An inshore shark found in continental and insular shelves from less than a metre to at least 170 m depth. Feeds on small fishes, cephalopods and crustaceans. Viviparous, with yolk-sac placenta; producing 1 or 2 pups per litter; gestation period of about 12 months every two years.

Conservation status: IUCN Red List: Near Threatened (assessed 2003).

Remarks: The only whaler shark which has an extremely calcified rostral cartilage that can be felt by pinching the snout forward of the eyes; most obvious in adult specimens. Known to form large, sexually segregated schools. Often confused with *Rhizoprionodon* species, but differs in having the second dorsal-fin origin about level with mid-base of anal fin (vs. in level with anal-fin insertion). Sometimes confused with juvenile Spinner Sharks (p. 50) but differs in having a much longer first dorsal-fin free rear tip.

References: None

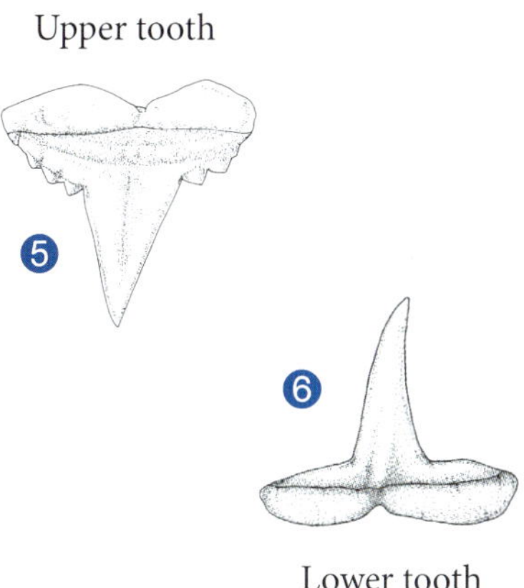

Image details: Lateral and ventral head: Dubai, United Arab Emirates (adult male ~75 cm TL, 11th Oct 2012).

Carcharhinidae (Whaler Sharks)

Blacktip Reef Shark

Carcharhinus melanopterus (Quoy & Gaimard, 1824)

Ventral head

<table>
<tr><td rowspan="3" style="writing-mode: vertical-lr">SIZE</td><td>Maximum: 180 cm TL, but mostly less than 160 cm TL</td></tr>
<tr><td>Maturity: males at 91–100 cm TL; females at 96–112 cm TL</td></tr>
<tr><td>At birth: 33–52 cm TL</td></tr>
</table>

KEY FEATURES

❶ First dorsal fin with a broad black tip, often with a whitish area beneath

❷ Snout very short and broadly rounded

❸ No interdorsal ridge

❹ Yellowish brown with a distinct pale stripe on each side

❺ Lower lobe of caudal fin with a broad black tip

❻ Upper teeth narrowly triangular, oblique, with coarse basal serrations

❼ Lower teeth narrow, more erect and finely serrated

Carcharhinidae (Whaler Sharks)

Worldwide distribution: Indo-West and Central Pacific, from southeastern Africa to the Central Pacific Islands; also in the eastern Mediterranean by invasion through Suez Canal.

Gulf occurrence: Patchy and apparently uncommon; confirmed from the western United Arab Emirates and a single record off Kuwait.

Habitat & biology: A coastal species found over coral reefs and tidal flats in depths of less than a metre to at least 75 m. Feeds on small bony fishes, cephalopods, crustaceans, seasnakes and seabirds. Viviparous, with yolk-sac placenta; producing 2–4 pups per litter. A 10 month gestation period was recorded for this species off Moorea in French Polynesia and 11 months of northern Australia.

Conservation status: IUCN Red List: Near Threatened (assessed 2009).

Remarks: Usually one of the most common coral reef sharks, but not commonly seen on Gulf reefs. Important for aquarium display and tourism in many parts of its range. Males and females mature at 4.2 and 8.5 years, respectively, in northern Australia.

References: Porcher (2005); Moore *et al.* (2012b); Chin *et al.* (2013)

Blacktip Reef Sharks are often seen swimming in very shallow water, as here on Sir Bani Yas Island, United Arab Emirates, January 2010

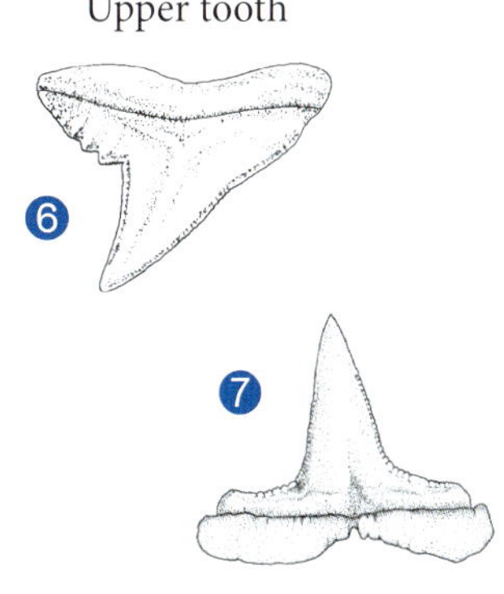

Image details: Lateral: Abu Dhabi, United Arab Emirates (~62 cm TL, 9th Apr 2010); Ventral head: Dubai, United Arab Emirates (24th May 2012).

Sandbar Shark

Carcharhinus plumbeus (Nardo, 1827)

Ventral head

<table>
<tr><td rowspan="1">SIZE</td><td>

Maximum: at least 239 cm TL

Maturity: males at 131–178 cm TL; females at 144–183 cm TL

At birth: 56–75 cm TL

</td></tr>
</table>

KEY FEATURES

❶ First dorsal fin very tall and broad, its origin about level with pectoral-fin insertions

❷ Snout moderately long and broadly rounded

❸ Interdorsal ridge present

❹ No distinct fin markings

❺ Upper teeth broadly triangular, slight oblique and serrated

❻ Lower teeth narrow, more erect and finely serrated

Worldwide distribution: Cosmopolitan with a patchy distribution in all warm temperate and tropical waters.

Gulf occurrence: Poorly known; recently confirmed from shark landings in the United Arab Emirates.

Habitat & biology: A coastal species found in estuaries and bays, including intertidal regions, to at least 280 m depth. Forages near the seabed feeding on bottom fishes, small sharks, rays, and crustaceans. Viviparous, with yolk-sac placenta; producing 1–14 pups per litter; gestation period is about 8–12 months off Florida (USA) and off Western Australia.

Conservation status: IUCN Red List: Vulnerable (assessed 2009).

Remarks: In Western Australian waters, females and males mature at 16 and 14 years and reach 25 and 19 years of age, respectively. Severe population declines have been reported in the North Atlantic.

References: Cliff *et al.* (1988); McAuley *et al.* (2006); Jabado *et al.* (2014)

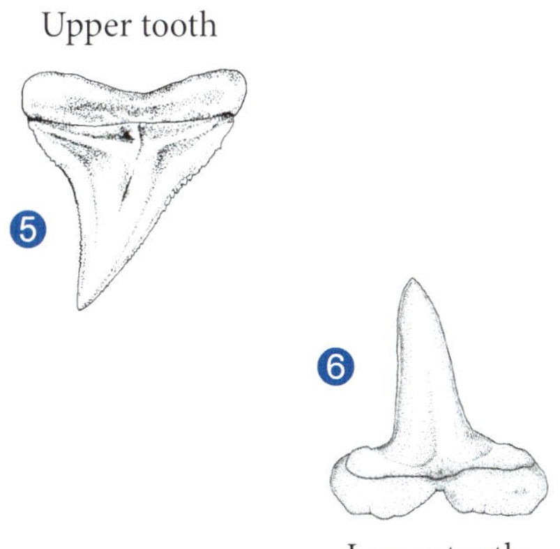

Image details: Lateral: Dubai, United Arab Emirates (female ~120 cm TL, 18th Dec 2011); Ventral head: Indonesia (juvenile male 87.5 cm TL, 28th May 2005).

Spot-tail Shark

Carcharhinus sorrah (Müller & Henle, 1839)

Ventral head

SIZE

Maximum: 166 cm TL

Maturity: males at 93–108 cm TL; females at 95–118 cm TL

At birth: 45–60 cm TL

KEY FEATURES

❶ Second dorsal, pectoral and lower caudal fin with distinct black tips

❷ Interdorsal ridge present

❸ Snout moderately long and pointed

❹ First dorsal-fin origin about opposite pectoral-fin free rear tips

❺ Second dorsal fin very long, its inner margin very long

❻ Upper teeth triangular, oblique, lateral margin notched

❼ Lower teeth similar to uppers but narrower

Worldwide distribution: Indo-West Pacific, from southeastern Africa to northern Australia and southern Japan.

Gulf occurrence: Widespread and common.

Habitat & biology: An inshore species found from the intertidal region to at least 140 m depth. Feeds on bony fishes, cephalopods and crustaceans. Viviparous, with yolk-sac placenta; producing 1–8 pups per litter. Gestation period 9–10 months in northern Australian populations.

Conservation status: IUCN Red List: Near Threatened (assessed 2009).

Remarks: Juveniles of this species are commonly seen in fish markets in the Gulf region. In northern Australian waters, females and males mature at 2.3–2.4 years and live to at least 14 years of age.

References: Moore *et al.* (2012a); Harry *et al.* (2013)

Carcharhinidae (Whaler Sharks)

Finned Spot-tail Shark carcasses at the Doha fish
market in Qatar, April 2009

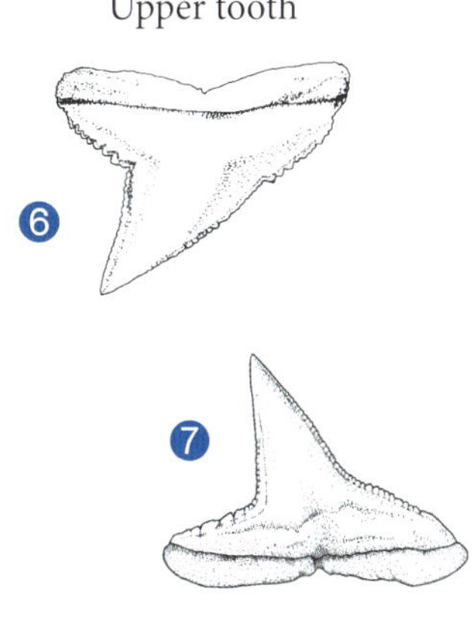

Image details: Lateral and ventral head: Dubai, United Arab Emirates (juvenile ~80 cm TL, 6th Oct 2012).

Tiger Shark

Galeocerdo cuvier (Péron & Lesueur, 1822)

Ventral head

<table>
<tr><td>SIZE</td><td>Maximum: 740 cm TL, mostly less than 500 cm TL
Maturity: males at 226–292 cm TL; females at 250–350 cm TL
At birth: 51–76 cm TL</td></tr>
</table>

KEY FEATURES

❶ A small, slit-like spiracle present

❷ Caudal peduncle with a low, rounded lateral keel

❸ Body with a pattern of dark, vertical bars (less distinct in adults)

❹ Upper labial furrows extremely long

❺ Snout very short and bluntly rounded

❻ Teeth in both jaws heavily serrated, cockscomb-shaped, deeply notched on one side

Worldwide distribution: Cosmopolitan in tropical waters; often making seasonal excursions into warm temperate waters.

Gulf occurrence: Probably widespread historically, but appears to be now rare.

Habitat & biology: A wide-ranging species that can be found from shallow inshore waters to the open ocean far offshore. Occurs from the intertidal region to at least 140 m depth. Feeds on a wide range of prey including turtles, fish, marine mammals (e.g. dugongs), seabirds, sea snakes, crustaceans, molluscs, sharks and rays. The only ovoviviparous carcharhinid (without yolk-sac placenta); producing 3–82 pups per litter. In Hawaiian populations, gestation lasts about 15–16 months and females give birth every three years.

Conservation status: IUCN Red List: Near Threatened (assessed 2009).

Remarks: Males and females mature at 7–8 and 7–12 years, respectively. It is considered to be one of the most dangerous sharks due to its indiscriminate feeding and occurrence in shallow water. Target of shark control programs in some countries. In Western Australia, they have been shown to be a key component of seagrass ecosystems, by regulating feeding and behaviour of prey such as dugongs.

References: Whitney & Crow (2007)

Tiger Shark, probably landed
in Bahrain in the 1970s

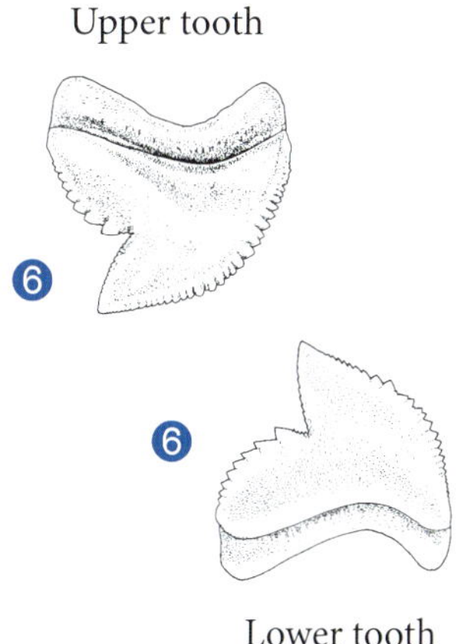

Image details: Lateral: Indonesia (female 193 cm TL, 17th Apr 2004); Ventral head: northern Australia (juvenile 2nd Feb 1993).

Sliteye Shark

Loxodon macrorhinus Müller & Henle, 1839

Ventral head

SIZE

Maximum: 99 cm TL

Maturity: males at 62–66 cm TL; females at 66–79 cm TL

At birth: 40–43 cm TL

KEY FEATURES

❶ Eye with a distinct notch on posterior margin

❷ Second dorsal fin much smaller than anal fin, its origin over anal-fin insertion

❸ First dorsal origin well posterior to pectoral-fin free rear tips

❹ Labial furrows small and inconspicuous

❺ Preanal ridges very long (equal to anal-fin base length)

❻ Snout very long and narrowly parabolic; almost translucent

❼ Teeth in both jaws with narrowly triangular, strongly oblique, smooth-edged cusps

Worldwide distribution: Indo-West Pacific, from southeastern Africa through to northern Australia and southern Japan.

Gulf occurrence: Widespread, but patchy; locally abundant in some areas.

Habitat & biology: A coastal species found near the bottom in shallow, clear waters in depths of 7 m to at least 100 m. Feeds on small bony fishes, crustaceans and cephalopods. Viviparous, with yolk-sac placenta; producing 1–4 pups per litter. In northeastern Australia, has an annual, seasonal reproductive cycle with a gestation period of 9–12 months.

Conservation status: IUCN Red List: Least Concern (assessed 2003).

Remarks: Molecular analyses suggest that the Western Indian Ocean population may be distinct from those in the Eastern Indian and West Pacific Oceans. In northeastern Australia, it was found to strongly associate with clear water, and males and females were found to mature at 1.9 and 1.4 years, respectively.

References: Gutteridge *et al.* (2011, 2013); Moore *et al.* (2012a); Naylor *et al.* (2012)

Upper tooth

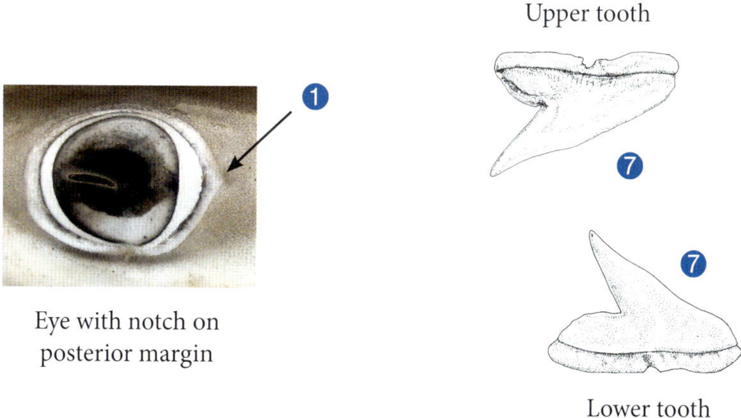

Eye with notch on
posterior margin

Lower tooth

Image details: Lateral and ventral head: Dubai, United Arab Emirates (adult male 72 cm TL, 9th Oct 2012); Eye: Dubai, United Arab Emirates (female 59 cm TL, 9th Oct 2012).

Sicklefin Lemon Shark

Negaprion acutidens (Rüppell, 1837)

Ventral head

<table>
<tr><td rowspan="3">SIZE</td><td>Maximum: 310 cm TL</td></tr>
<tr><td>Maturity: both sexes at 220–240 cm TL</td></tr>
<tr><td>At birth: 45–80 cm TL</td></tr>
</table>

<table>
<tr><td rowspan="6">KEY FEATURES</td><td>❶ Second dorsal fin similar in height to first dorsal fin</td></tr>
<tr><td>❷ Snout short and very broadly rounded</td></tr>
<tr><td>❸ No interdorsal ridge</td></tr>
<tr><td>❹ Anal fin nearly as large as second dorsal fin</td></tr>
<tr><td>❺ Teeth in both jaws slender, erect, smooth-edged</td></tr>
<tr><td>❻ Pale yellowish brown</td></tr>
</table>

Worldwide distribution: Indo-West and Central Pacific, from southeastern Africa through to the Central Pacific Islands.

Gulf occurrence: Rare; the only published record is from Jana Island (Saudi Arabia) in the 1970s. Further records are of interest.

Habitat & biology: A coastal species found in shallow waters around coral reefs and sand flats; prefers turbid waters of estuaries and lagoons; from the intertidal zone to at least 92 m depth. Feeds mainly on bony fishes, as well as cephalopods, crustaceans and rays. Viviparous, with yolk-sac placenta; producing 1–13 pups per litter. Off the Aldabra atoll in the Western Indian Ocean, gestation was found to last for 10–11 months.

Conservation status: IUCN Red List: Vulnerable (assessed 2003).

Remarks: Molecular studies revealed that it has a very restricted dispersal, thus is particularly vulnerable to overexploitation and has undergone severe declines in parts of its range.

References: Moore *et al.* (2010)

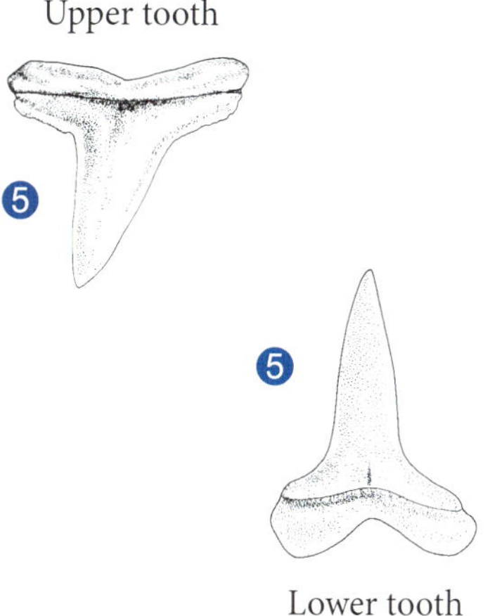

Image details: Lateral and ventral head: Almukalla fish market, Yemen (juvenile ~60 cm TL, 25th Mar 2013).

Milk Shark

Rhizoprionodon acutus (Rüppell, 1837)

Ventral head

KEY FEATURES

❶ Second dorsal fin much smaller than anal fin, its origin over anal-fin insertion

❷ No notch on posterior margin of eye

❸ Upper labial furrows long and prominent (1.4–2.0% of TL)

❹ Preanal ridges very long (equal to anal-fin base length)

❺ Enlarged pores alongside mouth corners in a distinct series (more than 16 in total)

❻ Snout long and narrowly rounded

❼ Teeth in both jaws with narrowly triangular, oblique, smooth-edged cusps

Worldwide distribution: Indo-West Pacific, from southeastern Africa through to northern Australia and southern Japan, and the Eastern Atlantic (West Africa).

Gulf occurrence: Widespread and abundant.

Habitat & biology: A primarily coastal species, found from the surface to 200 m depth. Feeds on small bottom-dwelling fishes, cephalopods and crustaceans. Viviparous, with yolk-sac placenta. In Oman, found to produce 1–8 pups per litter, with larger females generally having more pups; birth takes place in all seasons, but may peak in spring.

Conservation status: IUCN Red List: Least Concern (assessed 2003).

Remarks: Molecular analyses, along with regional differences in size and reproductive parameters, suggest this is possibly a species complex. One of the most common sharks in the Gulf. In northeastern Australia, males and females found to mature at 1.1 and 1.8 years of age, respectively. The common name is derived from an Indian belief that the consumption of its meat promotes lactation for breast-feeding.

References: Henderson *et al.* (2006); Harry *et al.* (2010); Moore *et al.* (2012a); Naylor *et al.* (2012)

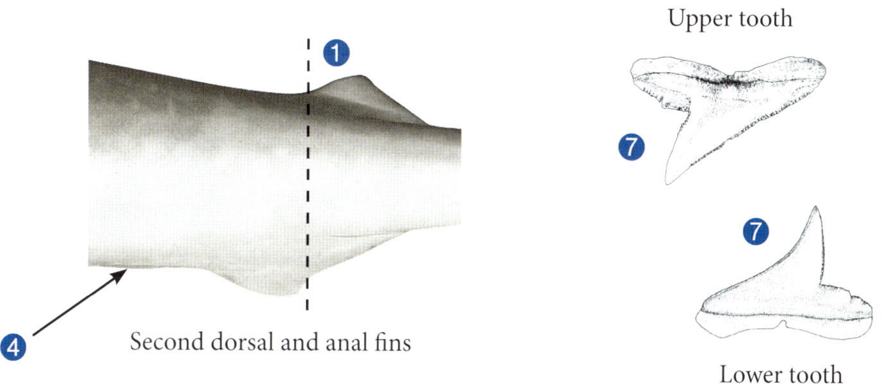

Image details: Lateral, ventral head and second dorsal and anal fins: Dubai, United Arab Emirates (adult male ~65 cm TL, 7ᵗʰ May 2012).

Grey Sharpnose Shark

Rhizoprionodon oligolinx Springer, 1964

Ventral head

Maximum: 85 cm TL

Maturity: males at 29–53 cm TL (50% mature at 53 cm TL in the Gulf); females at 32–41 cm TL (larger in the Gulf)

At birth: 21–28 cm TL

❶ Second dorsal fin much smaller than anal fin, its origin over anal-fin insertion

❷ No notch on posterior margin of eye

❸ Upper labial furrows short (0.2–1.3% of TL)

❹ Enlarged pores alongside mouth corners in a distinct series (less than 14 in total)

❺ Snout long and narrowly rounded

❻ Teeth in both jaws with narrowly triangular, oblique, smooth-edged cusps

❼ Small gap often visible at symphysis of upper and lower jaws when mouth closed

Worldwide distribution: Indo-West Pacific, from the Gulf to northern Australia and southern Japan.

Gulf occurrence: Patchy; common in surveys of fish markets in Kuwait and Bahrain, but not in Qatar.

Habitat & biology: An inshore species found on continental and insular shelves; from the surface to at least 36 m depth. Feeds on small fishes, crustaceans and cephalopods. Viviparous, with yolk-sac placenta; producing 3–5 pups per litter; a single female (67 cm TL) from Kuwait contained three near-term embryos (23–28 cm TL).

Conservation status: IUCN Red List: Least Concern (assessed 2003).

Remarks: As with other members of this genus, is highly productive and locally abundant. The small gap present at symphysis of upper and lower jaws (see feature 7 above) is good field-based character to use for this species which is often confused with the morphologically similar *Rhizoprionodon acutus*.

References: Moore *et al.* (2012a); Moore & Peirce (2013)

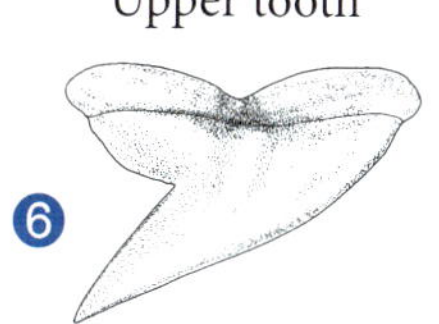
Upper tooth

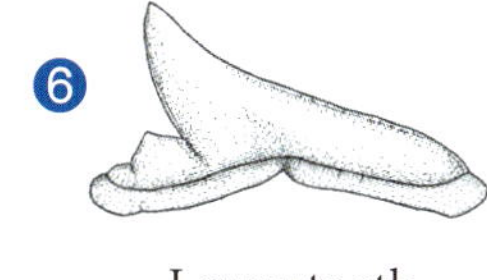
Lower tooth

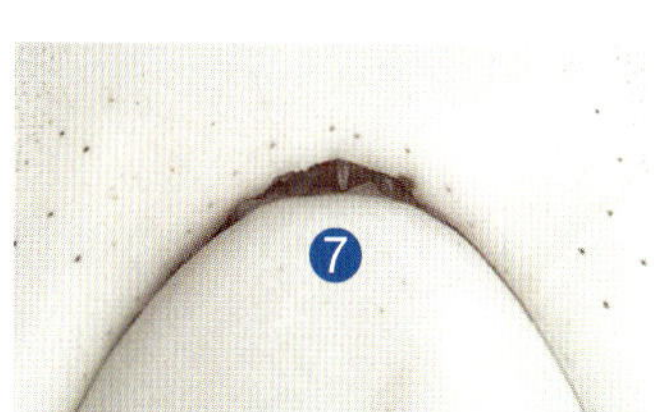
Gap visible at symphysis of upper and lower jaws

Image details: Lateral: Borneo, Malaysia (adult male 51 cm TL, 29th Apr 2004); Ventral head: Kuwait (female 70 cm TL, 4th Apr 2011); Mouth: Malaysia (adult male 51 cm TL, 29th Apr 2004).

Scalloped Hammerhead

Sphyrna lewini (Griffith & Smith, 1834)

Ventral head

Maximum: 420 cm TL

Maturity: males at 140–165 cm TL; females at 212–230 cm TL

At birth: 42–55 cm TL

KEY FEATURES

❶ Anterior margin of head curved, shallowly indented at midline

❷ First dorsal fin tall, moderately falcate

❸ Second dorsal fin low, with a very long free rear tip nearly reaching caudal fin

❹ Second dorsal-fin base about equal in length to anal-fin base

❺ Pelvic fins with straight posterior margins

❻ Upper teeth narrowly triangular, erect to oblique, smooth-edged

❼ Lower teeth narrower and more erect

Worldwide distribution: Cosmopolitan in all warm temperate and tropical seas.

Gulf occurrence: Possibly widespread, but not well recorded; may be more common in deeper waters of the easternmost Gulf.

Habitat & biology: An inshore and offshore species, found on continental and insular shelves, as well as far offshore; from the surface to at least 500 m depth; juveniles found more inshore, including enclosed bays while adults can be found well offshore, often around seamounts. Form large schools during the day but disperse into small groups or solitarily at night to feed. Feeds on bony fishes, sharks, rays and cephalopods. Viviparous, with yolk-sac placenta; producing litters of 14–41 pups after an 8–12 months gestation period.

Conservation status: IUCN Red List: Endangered (assessed 2007).

Remarks: The most common hammerhead species globally and an important component of many shark fisheries. Large adult aggregations are found in a number of places, e.g. Musandam off Oman and off Sudan in the Red Sea. Growth parameters vary between populations. Molecular studies have recently revealed that this is possibly a species complex, but no external diagnostic characters are currently available.

References: Abercrombie *et al.* (2005); Piercy *et al.* (2007); White *et al.* (2008); Naylor *et al.* (2012)

Scalloped hammerhead landings at Deira fish market, Dubai; caught by longlines operating in waters off Oman

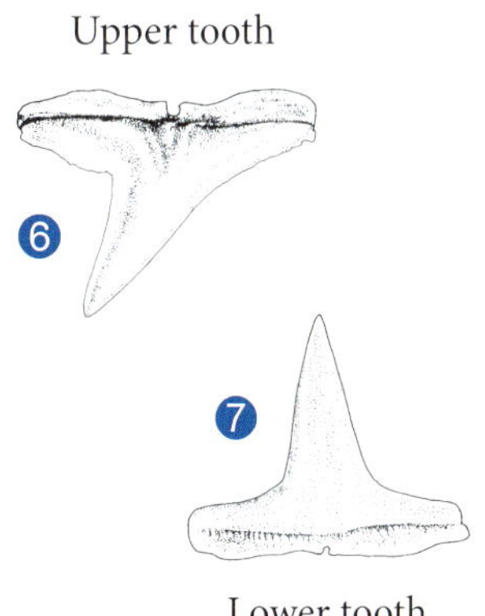

Image details: Lateral and ventral head: Dubai, United Arab Emirates (juvenile male ~130 cm TL, 22nd May 2012).

Sphyrnidae (Hammerhead Sharks)

Great Hammerhead

Sphyrna mokarran (Rüppell, 1837)

Ventral head

<table>
<tr><td rowspan="1">SIZE</td><td>

Maximum: 610 cm TL

Maturity: males at 234–269 cm TL; females at 250–300 cm TL

At birth: 50–70 cm TL

</td></tr>
<tr><td>KEY FEATURES</td><td>

❶ Anterior margin of head nearly straight, shallowly indented at midline

❷ First dorsal fin very tall, strongly falcate in adults

❸ Second dorsal fin moderately tall, with a short free rear tip

❹ Anal-fin base longer than second dorsal-fin base

❺ Pelvic fins with concave posterior margins

❻ Upper teeth triangular, oblique, with serrated edges

❼ Lower teeth narrower and slightly more erect

</td></tr>
</table>

Worldwide distribution: Cosmopolitan in all warm temperate and tropical seas.

Gulf occurrence: Widespread, but not common.

Habitat & biology: A wide-ranging species that can be found in very shallow, coastal waters from the intertidal region to at least 80 m depth; can also be found well offshore. Feeds on a range of prey particularly bottom-associated fishes such as stingrays, groupers and catfishes. Viviparous, with yolk-sac placenta; producing 6–42 pups per litter after a 7–11 month gestation period.

Conservation status: IUCN Red List: Endangered (assessed 2007).

Remarks: A very large (~500 cm TL) pregnant female was found dead on the shore of Kuwait Bay with its fins removed in April 2009. In northern Australia, age at maturity was at about 8 years, compared to 5–6 years in the Atlantic Ocean. Maximum age recorded is 44 years.

References: Stevens & Lyle (1989); Piercy *et al.* (2010); Harry *et al.* (2011)

Great Hammerheads on sale in Kuwait, 2008

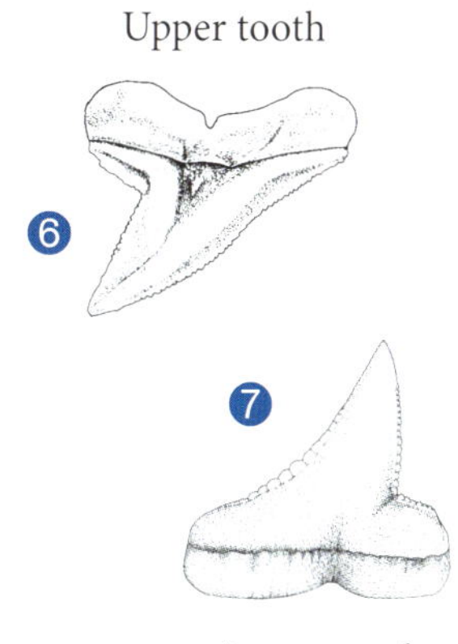

Image details: Lateral and ventral head: Dubai, United Arab Emirates (female ~150 cm TL, 24th May 2012).

Narrow Sawfish

Anoxypristis cuspidata (Latham, 1794)

Ventral head

<table>
<tr><td rowspan="3">SIZE</td><td>Maximum: 470 cm TL</td></tr>
<tr><td>Maturity: males at 200 cm TL; females at 225–230 cm TL</td></tr>
<tr><td>At birth: 43–70 cm TL</td></tr>
</table>

KEY FEATURES

❶ First dorsal-fin origin well posterior to pelvic-fin origins

❷ Ventral lobe of caudal fin long and prominent

❸ Rostral saw very long and narrow, with 18–30 pairs of lateral teeth

❹ No teeth at base of rostral saw

❺ Rostral teeth closer together near tip than at middle of saw

❻ Nostrils long and narrow, with small anterior nasal flaps

Worldwide distribution: Indo-West Pacific, from the Gulf through to northern Australia and New Guinea, and northwards to Japan.

Gulf occurrence: Extremely rare; historical records are limited to the coast of Iran.

Habitat & biology: An inshore benthic species found from the intertidal to around 40 m depth, including estuaries; juveniles mainly found in shallower, coastal areas in less than 10 m depth. Feeds on small fishes and cephalopods. Viviparous, with yolk-sac dependency; producing litters of 6–23 pups, with larger females containing more pups; an annual birth cycle.

Conservation status: IUCN Red List: Endangered (assessed 2012).

Remarks: Once abundant, all species of sawfishes have undergone catastrophic declines since cheap plastic gillnets became widely available, as the tooth-studded rostrum of sawfish becomes easily entangled. The last record of this species from the Gulf was in the early 1990s. All records of sawfishes should be reported.

References: Peverell (2005); Moore (2014)

Image details: Dorsal body & ventral head: Queensland, Australia (female ~140 cm TL, 26th Nov 1997).

Green Sawfish

Pristis zijsron Bleeker, 1851

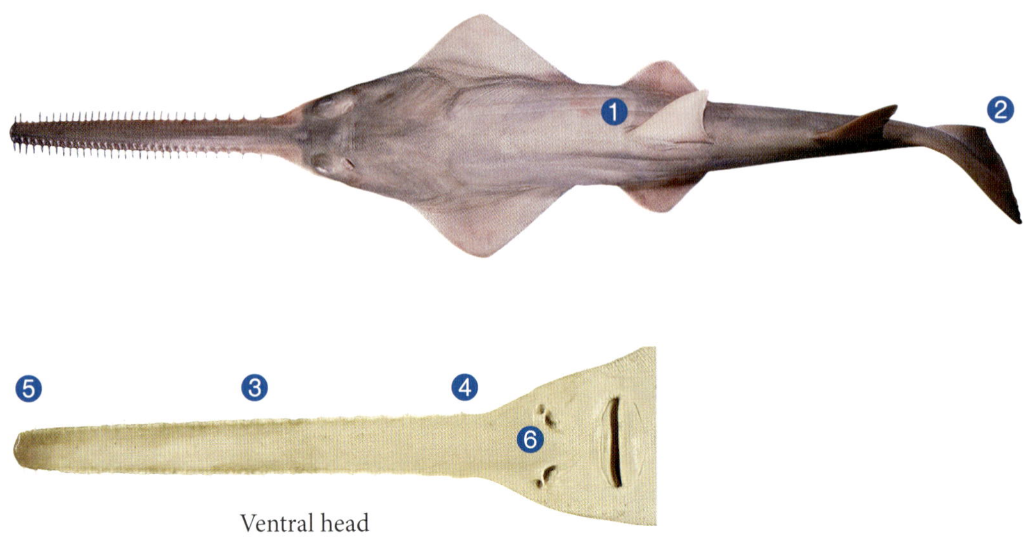

Ventral head

SIZE	Maximum: 540 cm TL, possibly to 730 cm TL Maturity: both sexes at about 300 cm TL At birth: 76–80 cm TL

KEY FEATURES

❶ First dorsal-fin origin slightly posterior to pelvic-fin origins

❷ Ventral caudal-fin lobe absent

❸ Rostral saw moderately long and slender, with 24–37 pairs of lateral teeth

❹ Rostral teeth present near saw base

❺ Rostral teeth much closer together near tip than at middle of saw

❻ Nostrils relatively broad, with a large anterior nasal flaps

Worldwide distribution: Indo-West Pacific, from South Africa to New Guinea and north to Vietnam.

Gulf occurrence: Extremely rare. Historically, the most widely distributed sawfish species in the Arabian region, with confirmed Gulf records from Kuwait to Abu Dhabi.

Habitat & biology: An inshore species, found in bays and estuaries, and also well offshore, over soft bottoms. Feeds on small schooling fishes and crustaceans. Viviparous, with yolk-sac dependency; producing litters of about 12 pups.

Conservation status: IUCN Red List: Critically Endangered (assessed 2012).

Remarks: Formerly abundant until around the 1960s, but recent fish surveys have failed to record this or other sawfish species in most of the Gulf region. Fully protected by no-take status in Bahrain and Qatar; all records of this or any sawfish should be reported. Occasional captures still occur in the central Gulf around Bahrain/Qatar, and into ports on the Musandam peninsula. The northern Gulf probably represents the northernmost limit of this tropical species, and anecdotal accounts from Kuwait suggest that sawfishes were seasonal migrants, occurring there in the summer months. Loss of intertidal areas from coastal development is likely to have negatively impacted sawfish species.

References: Stevens *et al.* (2005); Moore (2014)

Large sawfish were once common in the Gulf, as shown by
this old photograph from the middle part of the 20th C
(on display in the Bahrain National Museum, 2002)

Image details: Dorsal: Bahrain (91 cm TL, 20th Jun 1984); Ventral head: Queensland, Australia (juvenile male ~93 cm TL, 27th Nov 1991).

Pristidae (Sawfishes)

Bowmouth Guitarfish

Rhina ancylostoma Bloch & Schneider, 1801

SIZE

Maximum: 270 cm TL

Maturity: males at 150–178 cm TL; females at 180 cm TL

At birth: 45–51 cm TL

KEY FEATURES

❶ Head broadly rounded, well demarcated from pectoral fins

❷ First dorsal-fin origin just forward of pelvic-fin origin

❸ First dorsal fin larger than second dorsal fin

❹ Caudal fin large and lunate, with lower lobes more than half length of upper lobe

❺ Strong ridges bearing clusters of strong thorns on orbits, mid-body and shoulders

❻ Spiracles without skin folds on posterior margin

Worldwide distribution: Indo-West Pacific, from South Africa to New Guinea and north to southern Japan.

Gulf occurrence: Widespread, but uncommon; records include from off Iraq.

Habitat & biology: A benthic species found on coastal and offshore reefs; typically found on soft bottoms adjacent to coral reefs; in shallow water to depths of at least 90 m. Feeds primarily on crustaceans and molluscs. Viviparous, with yolk-sac dependency; producing litters of 2–11 pups, but biology poorly known.

Conservation status: IUCN Red List: Vulnerable (assessed 2003).

Remarks: Can be dangerous to handle when large due to the sharp, thorny ridges on head and body. Popular display animal for public aquaria. Fins are of high value and are targeted in some Asian countries leading to localised depletion in some areas. A female specimen caught locally in an intertidal fish trap (*hadrah*) was displayed in the aquarium of the Scientific Centre in Kuwait.

References: Devadoss & Batcha (1995); Raje (2006)

Ventral head

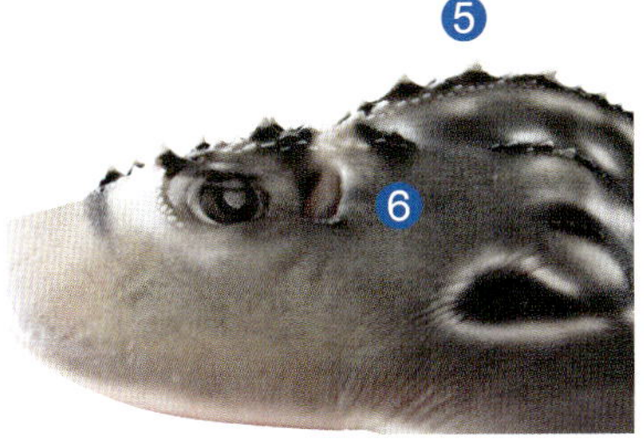

Lateral head

A large Bowmouth Guitarfish landed at Abu Dhabi (Mina Zayed Port, April 2010) being loaded into a van for overland transport

Image details: Dorsal body and ventral head: Dubai, United Arab Emirates (female ~130 cm TL, 24th Apr 2013); Lateral head: Indonesia (juvenile male 83 cm TL, 17th Apr 2004).

Whitespotted Wedgefish

Rhynchobatus djiddensis (Forsskål, 1775)

SIZE

Maximum: 310 cm TL

Maturity: not well known, probably over 150 cm TL

At birth: 60 cm TL

KEY FEATURES

❶ Origin of first dorsal-fin about over pelvic-fin origin

❷ Snout triangular, usually without a black blotch on ventral surface

❸ Spiracles with two skin folds on posterior margin

❹ A large black ring on each pectoral fin surrounded by about 4 small white spots

❺ Dark bars present between eyes

❻ Caudal fin with a distinct, well-developed lower lobe

Worldwide distribution: Not well defined; Western Indian Ocean, possibly from South Africa to the Gulf.

Gulf occurrence: Possibly widespread but not well defined; confirmed records from off Bahrain.

Habitat & biology: Usually found inshore, on soft bottoms and adjacent to reefs; from less than a metre to at least 70 m depth. Feeds on small fishes, cephalopods, bivalve molluscs and crustaceans. Viviparous, with yolk-sac dependency; litters of at least 4 pups, possibly many more in large females.

Conservation status: IUCN Red List: Vulnerable (assessed 2006).

Remarks: Previously considered to be widespread in the Indo-West Pacific, but recent taxonomic investigation has shown that there are multiple species, some of which overlap in at least part of their distribution. The closely-related, Indo-West Pacific species *Rhynchobatus australiae* and *R. laevis* are easily confused with this species. Fins are of very high value and localised depletions due to overfishing have been recorded in some areas.

References: Dudley & Cavanagh (2006); Moore (2012)

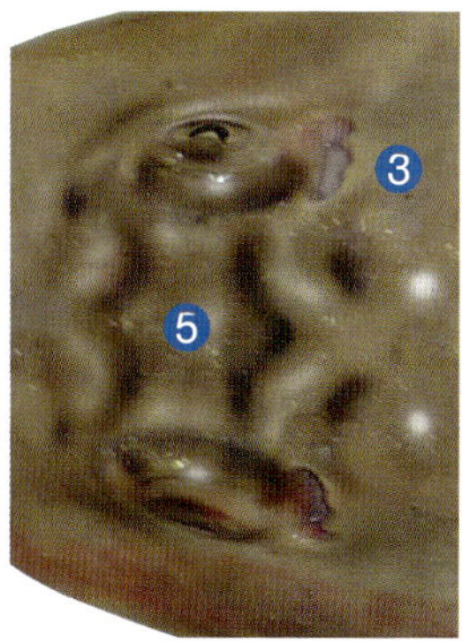

Eyes and spiracles

Adult Whitespotted Wedgefish landed in Oman

Image details: Dorsal and eyes and spiracles: Bahrain (~80 cm TL, 12th Apr 2012).

Smoothnose Wedgefish

Rhynchobatus laevis (Bloch & Schneider, 1801)

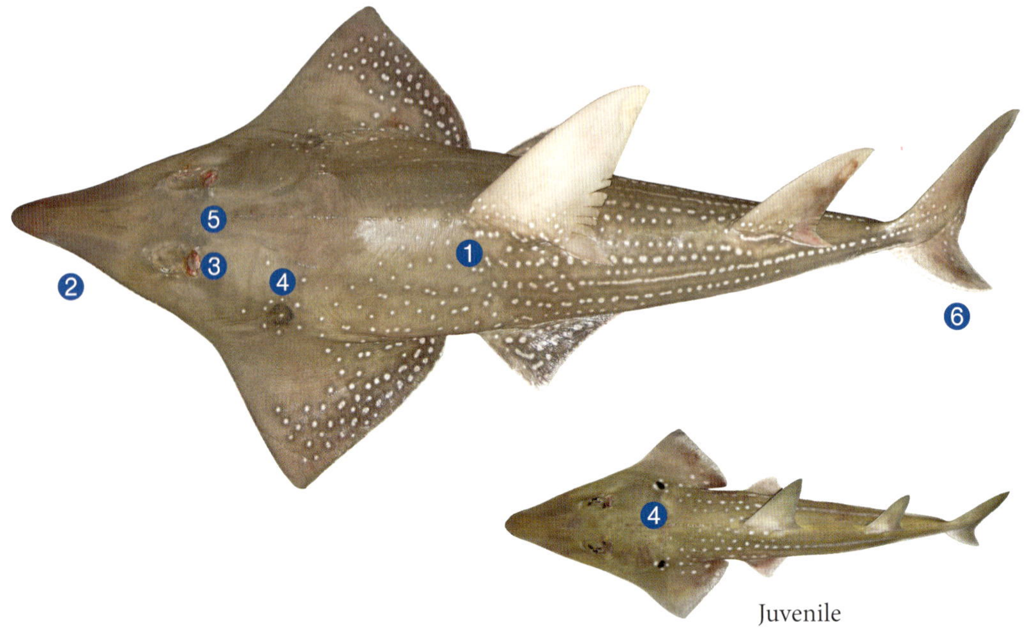

Juvenile

<table>
<tr><td>SIZE</td><td>Maximum: at least 270 cm TL

Maturity: unknown

At birth: probably around 50 cm TL</td></tr>
</table>

KEY FEATURES

❶ Origin of first dorsal-fin about over pelvic-fin origin

❷ Snout triangular, often with an irregular black blotch on ventral surface

❸ Spiracles with two skin folds on posterior margin

❹ A large black spot on each pectoral fin (sometimes forming a ring in large adults) surrounded by small white spots

❺ No dark bars between eyes

❻ Caudal fin with a distinct, well-developed lower lobe

Worldwide distribution: Not well defined; Indo-West Pacific, from South Africa to New Caledonia and north to Japan.

Gulf occurrence: Widespread; frequently occurs in small numbers in landings.

Habitat & biology: Usually found inshore, on soft bottoms and adjacent to reefs; from the intertidal zone to at least 60 m depth. Feeds mainly on crustaceans and molluscs, but also small fishes. Viviparous, with yolk-sac dependency; litters of at least 7–26 (average 15) pups after an unknown gestation period; non-seasonal reproductive cycle.

Conservation status: IUCN Red List: Vulnerable (assessed 2003).

Remarks: The taxonomy of this group is complicated with differences between some species very subtle and poorly defined. Similar to the Whitespotted Wedgefish *Rhynchobatus djiddensis* (p. 92) but can be distinguished by the absence of a dark interorbital bar which is present in the latter species. Fins are of very high value and localised depletions due to overfishing have been recorded in some areas.

References: Last & Stevens (2009)

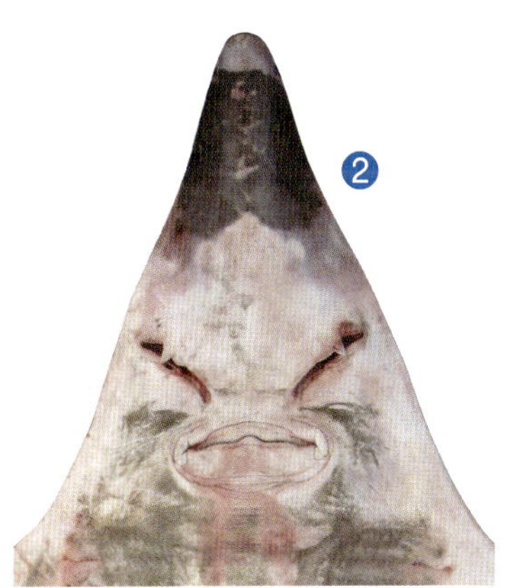

Ventral head

Wedgefish flesh for sale at Sharq market in Kuwait, 2008

Image details: Dorsal: Dubai, United Arab Emirates: Dorsal (adult female ~250 cm TL, 7th Oct 2012); Juvenile: Dubai, United Arab Emirates (juvenile male ~50 cm TL, 11th Oct 2012); Ventral head: Dubai, United Arab Emirates (female ~260 cm TL, 8th Oct 2012).

Sharpnose Guitarfish

Glaucostegus granulatus (Cuvier, 1829)

<table>
<tr><td rowspan="3">SIZE</td><td>Maximum: 215 cm TL (175 cm TL in the Gulf)</td></tr>
<tr><td>Maturity: unknown</td></tr>
<tr><td>At birth: unknown; smallest reported free-swimming individuals in the Gulf were 39 cm TL</td></tr>
</table>

KEY FEATURES

❶ Anterior nasal aperture almost rectangular, very large

❷ Snout long and slender anteriorly, eye length 9–14 times in distance from snout tip to eye

❸ Nostril width 1.0–1.2 times space between nostrils

❹ Posterior margin of spiracles with a weakly-developed outer fold and a rudimentary or absent inner fold

❺ Body plain, without spots or blotches

❻ Caudal fin without a distinct lower lobe

Worldwide distribution: Indo-West Pacific, from the Gulf to possibly New Guinea and northwards to China; not well defined.

Gulf occurrence: Patchy, but can be common where it occurs (e.g. Kuwait).

Habitat & biology: A benthic species that prefers soft substrates in coastal areas to depths of 119 m. Probably the species responsible for anecdotal accounts of large numbers of guitarfish feeding, with fins exposed, on an incoming tide on the intertidal flats of northern Kuwait. Viviparous, with yolk-sac dependency; producing litters of 6–10 pups; possibly with an annual reproductive cycle as in other rhinobatid species.

Conservation status: IUCN Red List: Vulnerable (assessed 2006).

Remarks: Often confused with *Glaucostegus typus* which differs from this species in having much larger nostrils (nostril width 2.2–2.4 times space between nostrils vs. 1.0–1.5 times) and a shorter and more anteriorly broader snout. Previously placed in the genus *Rhinobatos* but recent taxonomic studies elevated the subgenus *Glaucostegus* to generic level for this group of large guitarfishes.

References: Moore *et al.* (2012a)

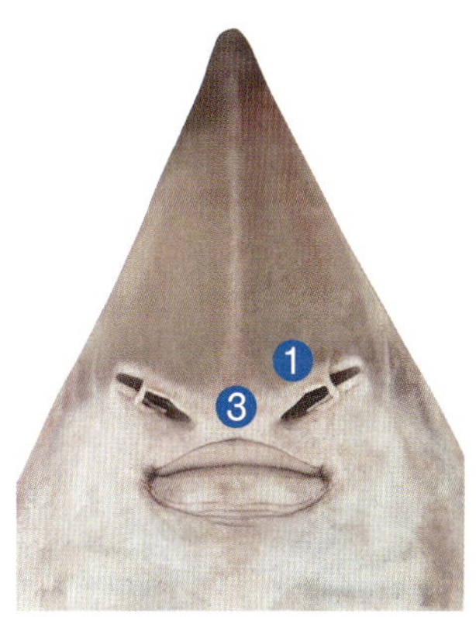

Ventral head

Large Sharpnose Guitarfish being sold at
Sharq fish market, Kuwait (April 2011)

Image details: Dorsal: Kuwait (39 cm TL, 21st Aug 1985); Ventral head: the Gulf (male ~100 cm TL).

Halavi Guitarfish

Glaucostegus halavi (Forsskål, 1775)

SIZE

Maximum: 171 cm TL

Maturity: males at 83 cm TL; females unknown

At birth: 29 cm TL

KEY FEATURES

❶ Anterior nasal aperture almost rectangular, very large

❷ Snout moderately long and broadly angular, eye length 4–6 times in distance from snout tip to eye

❸ Nostril width 1.3–1.5 times space between nostrils

❹ Posterior margin of spiracles with an outer and inner fold

❺ Body plain, without spots or blotches

❻ Caudal fin without a distinct lower lobe

Worldwide distribution: Western Indian Ocean, from the Red Sea to the Gulf; at least one confirmed record from the Mediterranean by invasion from the Red Sea through the Suez Canal. Records further westwards are unconfirmed.

Gulf occurrence: Patchy and uncommon; reported from fish markets in Bahrain and Abu Dhabi.

Habitat & biology: A coastal benthic species, found on soft bottoms in less than a metre to at least 40 m depth, but most commonly found in sheltered bays in less than 5 m depth. Feeds primarily on crustaceans, but probably also small fishes. Viviparous, with yolk-sac dependency; producing litters of up to 10 pups; pregnant females reported to move to shallow water to give birth.

Conservation status: IUCN Red List: Data Deficient (assessed 2008).

Remarks: Only recently confirmed from the Gulf. Juveniles probably utilise shallow water as a nursery area, thus coastal developments likely to impact heavily on this species.

References: Randall & Compagno (1995); Ben Souissi *et al.* (2007); Moore & Peirce (2013)

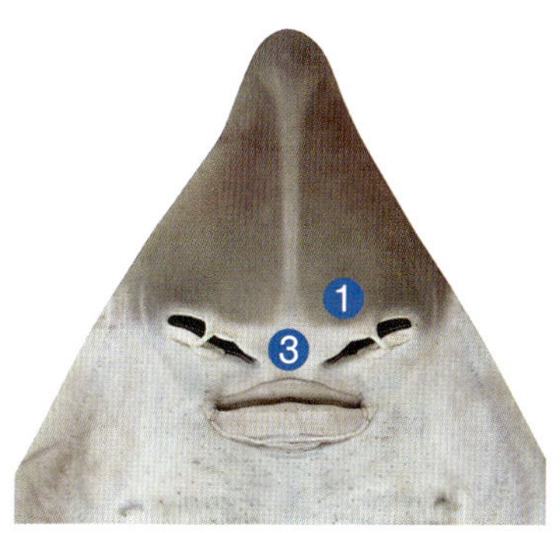

Ventral head

Large Halavi Guitarfish being sold at
Manama fish market, Bahrain (April 2012)

Image details: Dorsal: Oman, Arabian Sea (adult male 99.5 cm TL, 11th Nov 1993); Ventral head: the Gulf (adolescent male ~70 cm TL).

Arabian Spotted Guitarfish

Rhinobatos sp.

<table>
<tr><td>SIZE</td><td>Maximum: 79 cm TL
Maturity: unknown; 62.5 cm TL male paratype was mature
At birth: unknown</td></tr>
</table>

KEY FEATURES

❶ Anterior nasal aperture almost circular, relatively small

❷ Snout moderately long and relatively broad

❸ Anterior nasal flaps of nostrils not extending inwards and not nearly meeting at midline of internarial space

❹ Rostral cartilages not very close together

❺ Spiracles with two distinct skin folds

❻ Body with widely scattered white spots or ocelli (about size of pupil)

❼ Caudal fin without a distinct lower lobe

Worldwide distribution: Western Indian Ocean, known from the Gulf of Oman and the Gulf.

Gulf occurrence: Widespread, but not common.

Habitat & biology: Probably occurs on soft bottoms on the coastal and continental shelf; depth range unknown. Presumably feeds on small crustaceans as with other similar-sized guitarfish. Probably viviparous, with yolk-sac dependency; biology unknown.

Conservation status: IUCN Red List: Not Evaluated.

Remarks: Previously thought to be conspecific with *Rhinobatos punctifer*, but recently described as a new species. Colour pattern found to be highly variable in Gulf of Oman populations. Two other similar species from the Gulf of Oman could possibly also occur in the Gulf.

References: None

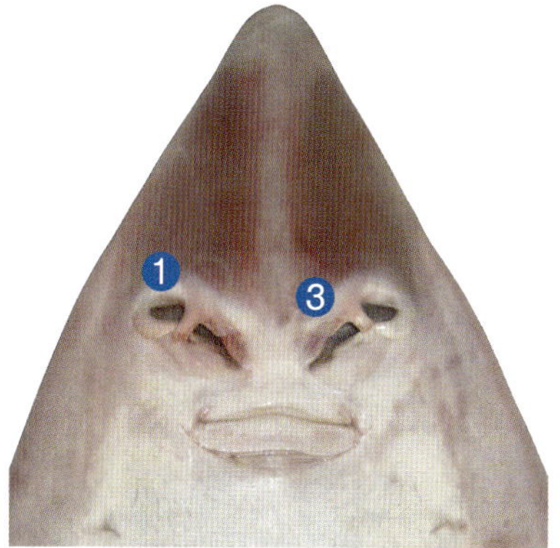

Ventral head

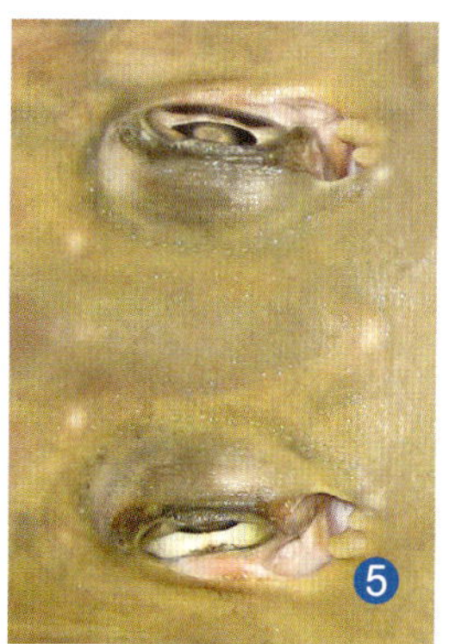

Dorsal view of eyes
and spiracles

Image details: Dubai, United Arab Emirates (adult male 64 cm TL, 10th Oct 2012).

Marbled Electric Ray

Torpedo sinuspersici Olfers, 1831

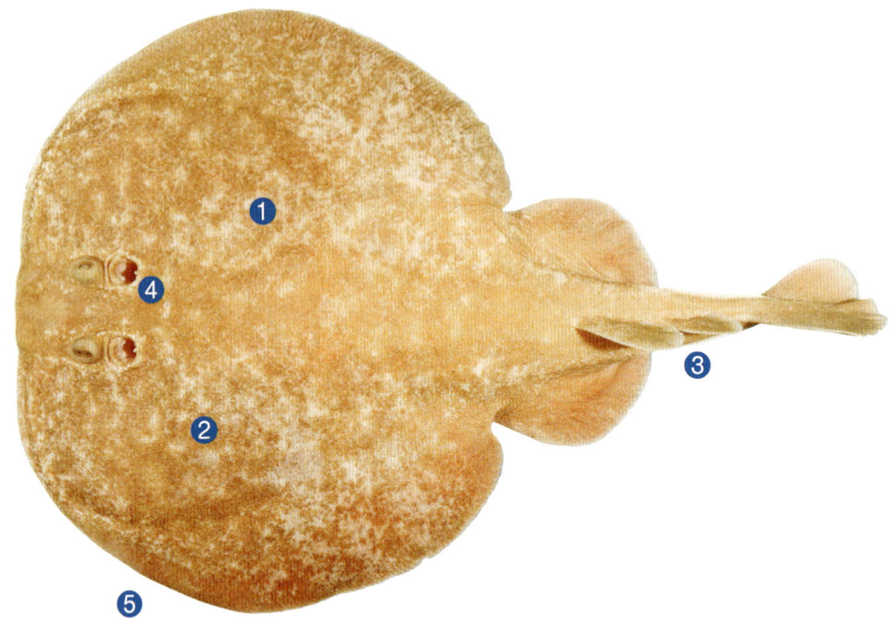

SIZE

Maximum: 130 cm TL, mostly less than 100 cm TL

Maturity: females by 63 cm TL

At birth: smallest free-swimming individual ~13 cm TL

KEY FEATURES

❶ Dorsal surface with broad cream or whitish vermiculations

❷ Numerous irregular, cream spots (not larger than eye) on anterior and lateral disc

❸ Dorsal fins very close together, space between them very short

❹ Small, knob-like papillae around margins of spiracles

❺ Disc flabby, with a large electric organ on each side

Worldwide distribution: Western Indian Ocean, from South Africa and Madagascar to the Gulf, and east to the Andaman Sea.

Gulf occurrence: Patchy; reported as rare in Kuwait in the 1960s, but relatively common in the central southern Gulf, especially below 50 m depth, in the 1980s.

Habitat & biology: A coastal species found on sandy and coral reef areas; from the surf zone to well offshore in depths to at least 200 m. Biology largely unknown; viviparous; three pregnant females off South Africa contained litters of 9, 15 and 22 pups.

Conservation status: IUCN Red List: Data Deficient (assessed 2006).

Remarks: Four *Torpedo* species are found in the Western Indian Ocean, with *T. sinuspersici* the only species originally described from Gulf waters. Further taxonomic revision of this complex of species is required to confirm the ranges of each species; although reported to have overlapping ranges, there are no confirmed records of them co-occurring in any one region. In 17th C Persia, electric rays were recorded as being abundant in net catches in the winter months. Colour can be variable but is the best method of identifying this species from its closest relatives. The image used to depict this species is of a juvenile specimen from Oman originally identified as *Torpedo panthera*; its colour pattern better matches that provided for *Torpedo sinuspersici*; further verification of the identity of this specimen is required.

References: Wallace (1967); Kuronuma & Abe (1972); Carruba & Bowers (1982); Carvalho *et al.* (2002b)

Image details: Dorsal: Masirah Island, Oman (15.5 cm TL, 19th Nov 1993).

Smalleye Stingray

Dasyatis microps (Annandale, 1908)

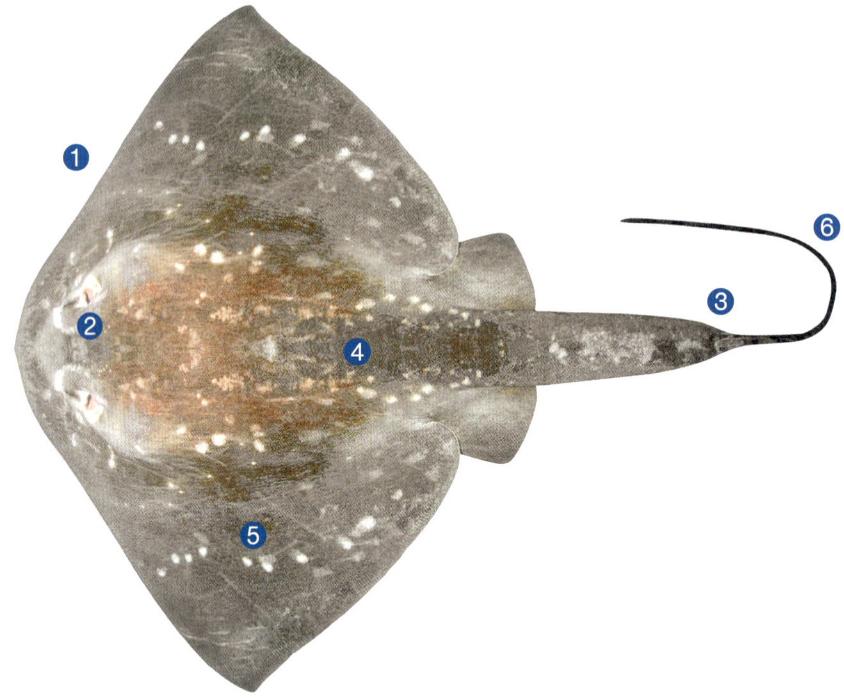

SIZE

Maximum: 222 cm DW (~320 cm TL)

Maturity: males by 158 cm DW; females by 180 cm DW

At birth: 31–33 cm DW

KEY FEATURES

❶ Disc very broad and strongly rhomboidal

❷ Eyes very small and spiracles large

❸ Tail with a very broad and depressed base, tapering markedly near sting

❹ No enlarged thorns along central disc or tail

❺ Brown to pinkish brown above with a diagonal row of white spots on each side

❻ Tail almost black beyond sting

Worldwide distribution: Patchy distributed in the Indo-West Pacific, from Mozambique to northern Australia and New Guinea.

Gulf occurrence: Patchy and rare; the only known records are from the central Gulf in the 1930s.

Habitat & biology: A coastal and probably semi-pelagic species, found in estuaries and over inshore reefs to depths of at least 50 m. Very little information exists for this species with only patchy records throughout its range. Viviparous, with histotrophy; one pregnant female recorded off India contained a single, full-term pup.

Conservation status: IUCN Red List: Data Deficient (assessed 2007).

Remarks: The broad disc of this species is unique amongst the dasyatid rays and is closer to the butterfly rays (Gymnuridae). This body shape is ideal for constant swimming, suggesting a semi-pelagic behaviour which is supported by observations of this species in mid-water over reefs in Mozambique. Dorsal coloration typically pale brown when observed underwater or when immediately taken from the water, but tends to become pinkish after landing.

References: Nair & Soundararajan (1976); Pierce *et al.* (2008); Moore (2010)

Underwater image of an adult Smalleye
Stingray off Mozambique

Image details: Dorsal: Mozambique (>1 m DW, 27th Mar 2012).

Pink Whipray

Himantura fai Jordan & Seale, 1906

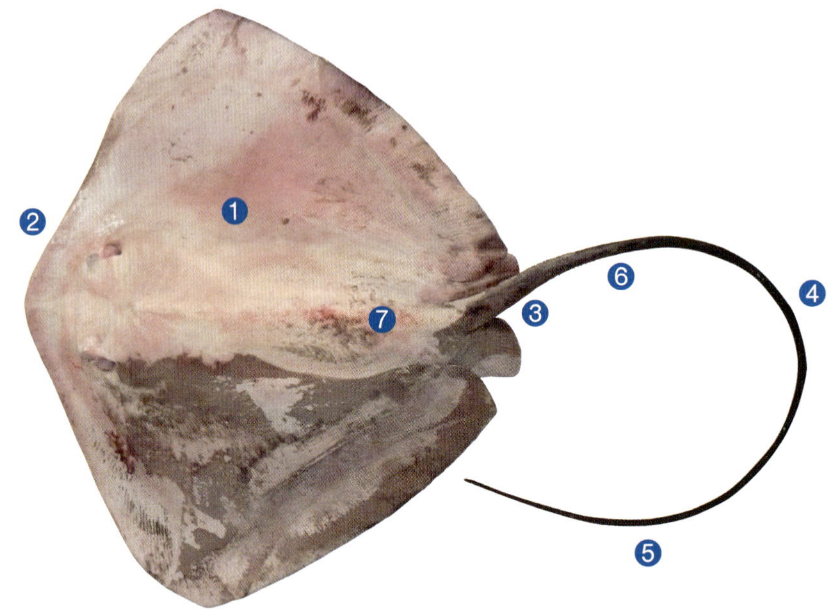

SIZE

Maximum: 184 cm DW (at least 500 cm TL)

Maturity: males at 108–122 cm DW; female size at maturity unknown

At birth: 30–60 cm DW (154–165 cm TL)

KEY FEATURES

❶ Upper surface uniformly greyish pink

❷ Disc quadrangular with a short, broad snout

❸ Tail base rounded in cross-section

❹ No skin folds on tail

❺ Tail very long, whip-like and uniformly dark beyond sting (not banded)

❻ Stinging spine located anteriorly on tail

❼ No enlarged thorny denticles on midline of disc and tail

Worldwide distribution: Probably widespread in Indo-West Pacific, from southern Africa to Micronesia.

Gulf occurrence: Not well documented; a single record exists from Abu Dhabi waters.

Habitat & biology: A primarily coastal species that prefers soft bottoms in bays and near reef habitats; from the intertidal region to at least 200 m depth, but mostly in less than 70 m depth; often forms small aggregations which can sit on top of each other and other large stingray species forming stacks. Feeds predominantly on shrimps and crabs. Viviparous, with histotrophy; litter size unknown.

Conservation status: IUCN Red List: Least Concern (assessed 2009).

Remarks: Overall distribution not well defined as often misidentified with the sympatric Jenkins' Whipray *Himantura jenkinsii* (p. 155) which differs in having enlarged thorny denticles on the midline of disc and tail.

References: White *et al.* (2006); Vaudo & Heithaus (2011); Moore (2012)

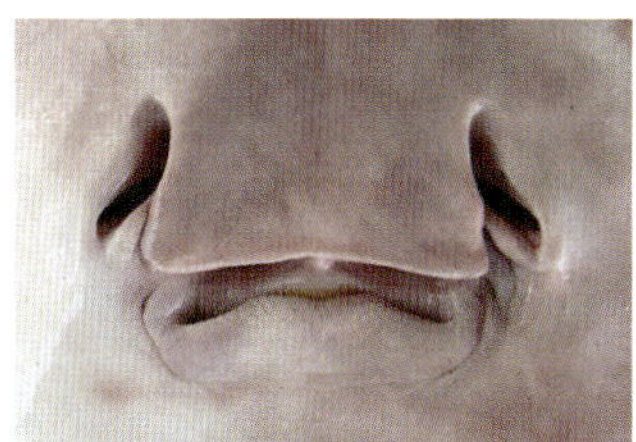

Oronasal region

Underwater image of two Pink Whiprays off Mauritius

Image details: Dorsal: Mina Zayed, Abu Dhabi, United Arab Emirates (female 124 cm DW, 11th Apr. 2010); Oronasal: Lombok, Indonesia (male embryo 45 cm DW, 3rd Aug. 2010).

Scaly Whipray

Himantura imbricata (Bloch & Schneider, 1801)

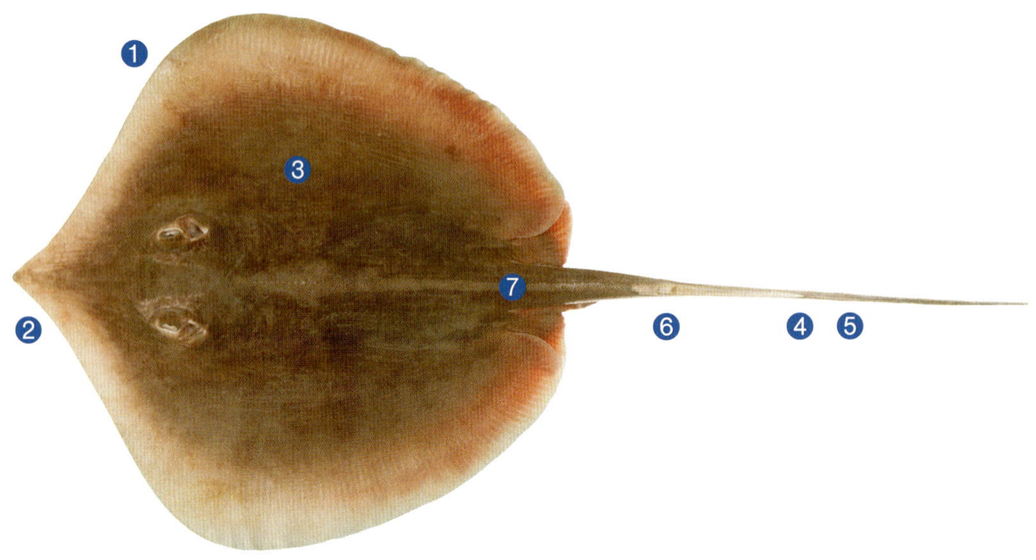

<table>
<tr><td rowspan="3">SIZE</td></tr>
</table>

SIZE	Maximum: 26 cm DW (up to 65 cm TL)
	Maturity: males at around 18 cm DW in the Gulf; females by at least 23 cm DW
	At birth: 10 cm DW

KEY FEATURES	❶ Profile of disc oval
	❷ Snout relatively long and pointed
	❸ Upper surface uniformly yellowish brown
	❹ Tail relatively short, not whip-like
	❺ No skin folds on tail
	❻ Sting situated anteriorly on tail
	❼ Enlarged thorns on midline of tail in adults

Worldwide distribution: Distribution poorly defined due to confusion with similar species; possibly restricted to the Western Indian Ocean from the Red Sea to India.

Gulf occurrence: Not well understood, but probably widespread; can be locally abundant.

Habitat & biology: A coastal species found on soft bottoms. Diet not known but probably consisting mainly of small benthic invertebrates. Biology poorly known; viviparous with histotrophy.

Conservation status: IUCN Red List: Data Deficient (assessed 2009).

Remarks: Previously considered to have a wider distribution to Thailand and Indonesia, but likely due to misidentification with the morphologically similar Dwarf Whipray *Himantura walga* and Java Whipray *Himantura javaensis* in those regions. Likely to be under-reported. Due to its small size, probably discarded by fishers. One of the most common elasmobranchs in shallow, turbid waters off northern Kuwait.

References: Last & White (2013); Bishop *et al.* (in press)

Underwater image of a specimen in coral habitat from
Qarou Island, Kuwait, 2011

Image details: Dorsal: Kuwait (male 35 cm TL, 21st Aug. 1985).

Dasyatidae (Stingrays)

Arabian Banded Whipray

Himantura randalli Last, Manjaji-Matsumoto & Moore, 2012

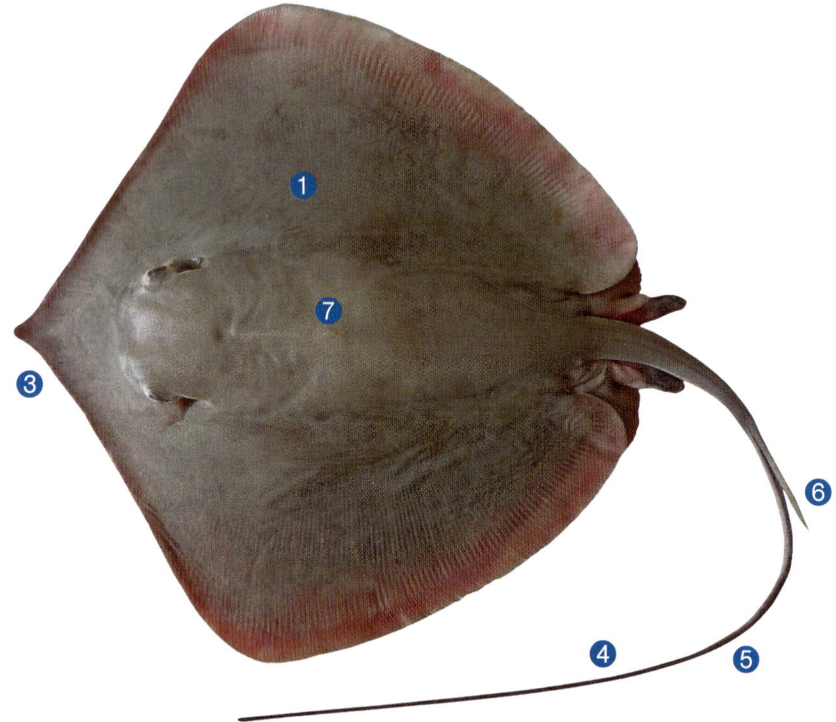

<table>
<tr><td rowspan="3">SIZE</td><td>Maximum: 62 cm DW</td></tr>
<tr><td>Maturity: males at 40 cm DW; female size at maturity unknown</td></tr>
<tr><td>At birth: 15–17 cm DW</td></tr>
</table>

KEY FEATURES

❶ Upper surface uniformly greenish grey

❷ Tail banded in juveniles (uniformly dark in adults)

❸ Disc quadrangular with a moderately long snout

❹ Tail very long, its base rounded in cross-section

❺ No skin folds on tail

❻ Sting situated anteriorly on tail

❼ Usually 1 or 2 enlarged denticles located on middle of disc

Worldwide distribution: Possibly endemic to the Gulf.

Gulf occurrence: Widespread; can be locally abundant.

Habitat & biology: A coastal species that prefers soft substrates such as sand and mud. Feeds primarily on shrimps and stomatopods. Presumably viviparous, with histotrophy, like other members of the genus.

Conservation status: IUCN Red List: Not evaluated.

Remarks: Belongs to the *Himantura gerrardi* complex, which is currently under taxonomic review. Previous records of *Himantura gerrardi* and *Dasyatis bennetti* from the Gulf are likely to be wholly or partly misidentifications of this species. Although often caught in local fisheries, of very limited value and often dumped.

References: Last *et al.* (2012)

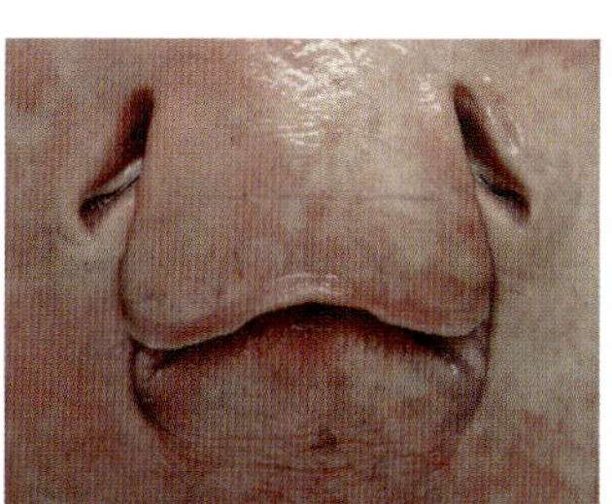

Oronasal region

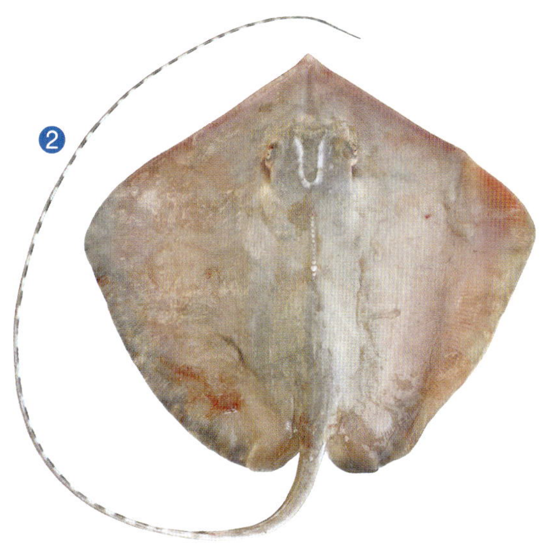

Juvenile

Image details: Dorsal and oronasal: Sharq fish market, Kuwait City, Kuwait (adult male holotype 41 cm DW, 5th Apr. 2011); Juvenile: Corniche fish market, Doha, Qatar (juvenile male 24 cm DW, 13th Apr. 2009).

Reticulate Whipray

Himantura uarnak (Gmelin, 1789)

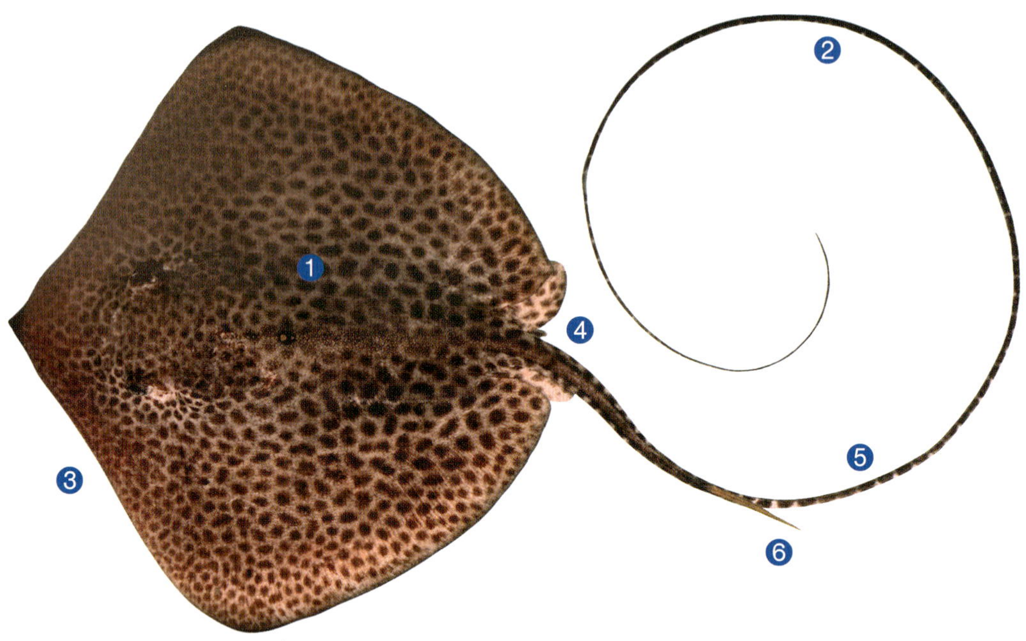

SIZE

Maximum: 160 cm DW (450 cm TL)

Maturity: males at 82–84 cm DW; females unknown

At birth: 21–28 cm (about 110 cm TL)

KEY FEATURES

❶ Upper surface strongly patterned with spots, reticulations or ocelli

❷ Tail very long and banded (banding less distinct in adults)

❸ Disc quadrangular

❹ Tail base rounded in cross-section

❺ No skin folds on tail

❻ Sting situated anteriorly on tail

Worldwide distribution: Widespread along the Indian and Western Pacific Oceans.

Gulf occurrence: Widespread; not particularly common.

Habitat & biology: Found on soft bottoms in coastal lagoons and estuaries and on the continental shelf; from the intertidal to at least 50 m depth. Feeds on crustaceans and small fishes. Viviparous, with histotrophy; producing litters of 4 pups.

Conservation status: IUCN Red List: Vulnerable (assessed 2009).

Remarks: Represents a complex of species in the Indo-West Pacific which need urgent taxonomic investigation; originally described from the Red Sea and considered to be widespread in the Indo-West Pacific, but the true *Himantura uarnak* may be restricted to the Indian Ocean.

References: Manjaji-Matsumoto & Last (2008)

Adult Reticulate Whipray with dense ocelli

Image details: Dorsal: Kuwait (juvenile male ~34 cm DW, 21st Aug 1985); Adult: Kuwait (adult male ~100 cm DW, 2nd Apr 2011).

Feathertail Stingray

Pastinachus ater (Macleay, 1883)

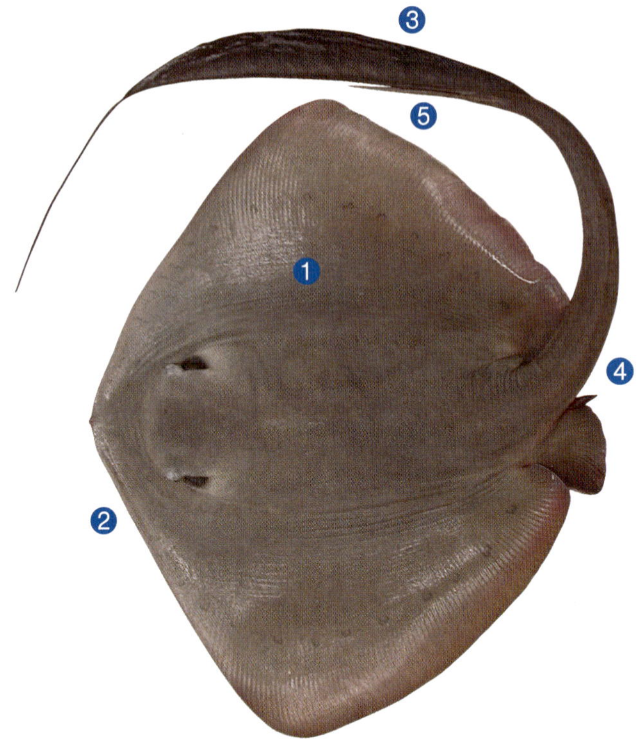

SIZE

Maximum: 200 cm DW (over 300 cm TL)

Maturity: males at 96–98 cm DW; female maturity unknown

At birth: 18 cm DW (50 cm TL)

KEY FEATURES

❶ Disc uniformly greyish black and quadrangular

❷ Snout short and rounded

❸ Ventral skin fold on tail very deep, blackish, and terminating well before tip of tail

❹ Tail base broad and depressed

❺ Stinging spine located posteriorly on tail (its distance from cloaca more than half disc width)

Worldwide distribution: Indo-West Pacific, from the Gulf through to Melanesia and Micronesia; Western Indian Ocean distribution not well defined.

Gulf occurrence: Possibly not common; confirmed from off Qatar.

Habitat & biology: A coastal species, also found in freshwater and estuaries, to depths of at least 60 m. Probably feeds on crustaceans and small fishes. Viviparous, with histotrophy; litters of two pups.

Conservation status: IUCN Red List: Not evaluated.

Remarks: Previously considered to be conspecific with the Cowtail Stingray *Pastinachus sephen* (p. 116) but recently resurrected as a valid species. Further investigation is required to confirm the validity of the scientific names of the two species occurring in the Gulf. In other regions, the common name of Cowtail Stingray is also attributed to this species.

References: White & Dharmadi (2007); Last & Manjaji-Matsumoto (2010)

A Feathertail or Cowtail Stingray on reef habitat off Kuwait

Image details: Dorsal: Queensland, Australia (juvenile male 64.5 cm DW, 31st May 2006).

Cowtail Stingray

Pastinachus sephen (Forsskål, 1775)

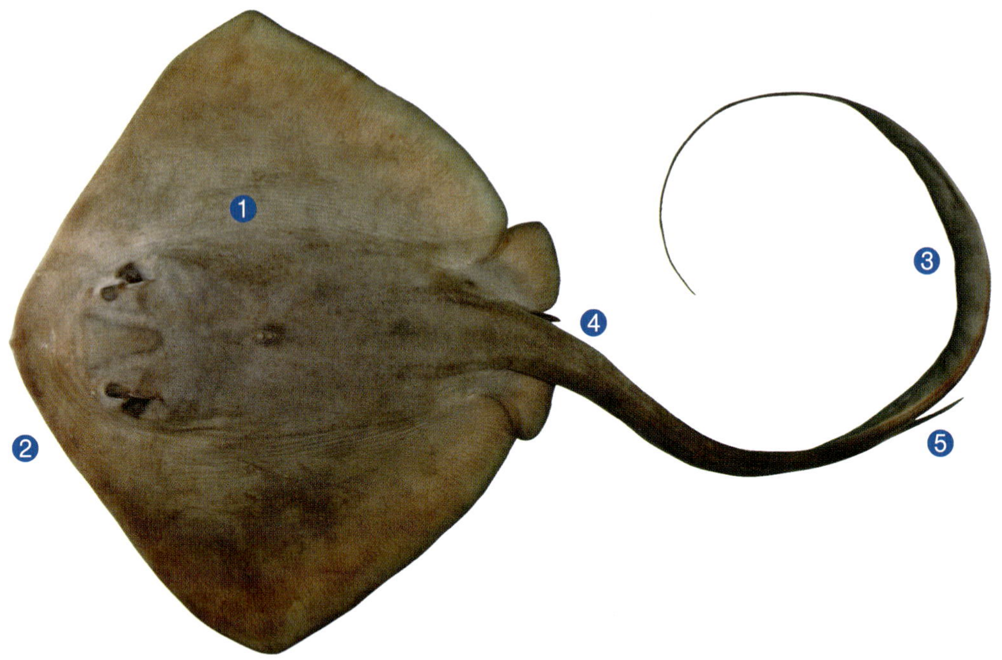

<table>
<tr><td rowspan="1">SIZE</td><td>

Maximum: 96 cm DW (in the Gulf)

Maturity: males by 54 cm DW and females by 61 cm DW in
        the Gulf

At birth: about 18 cm DW

</td></tr>
</table>

KEY FEATURES

❶ Disc uniformly greyish green and quadrangular

❷ Snout short and rounded

❸ Ventral skin fold on tail deep, greyish, and terminating
   well before tip of tail

❹ Tail base broad and depressed

❺ Stinging spine located posteriorly on tail (its distance from
   cloaca more than half disc width)

Worldwide distribution: Poorly defined; from South Africa to the Red Sea and the Gulf, and possibly to India.

Gulf occurrence: Not well defined; confirmed from off Kuwait; *Pastinachus* species frequently occur in bycatch.

Habitat & biology: A coastal species found in lagoons, mangroves, estuaries, coral and sandy habitat to at least 60 m depth. Feeds on crustaceans and small fishes. Viviparous, with histotrophy; produces a single pup per litter. Often landed as bycatch locally, but of little value and often dumped.

Conservation status: IUCN Red List: Data Deficient (assessed 2009).

Remarks: Previously considered to have a wide range in the Indo-West Pacific, but recent taxonomic studies revealed it is a species complex consisting of at least five species. Very similar to the Feathertail Stingray *Pastinachus ater* (p. 114) but differs in having a paler and slightly narrower ventral skin fold; further investigation is required to determine the key features separating these species; previous records of this species in the Gulf could refer to either species.

References: Last & Manjaji-Matsumoto (2010)

Cowtail rays are a frequent, but undesirable, bycatch of gillnet fisheries in the Gulf (Sharq market, Kuwait, 5th Apr 2011)

Image details: Dorsal: Kuwait Bay, Kuwait (juvenile male ~26 cm DW, 17th Aug 1985).

Blotched Fantail Ray

Taeniurops meyeni Müller & Henle, 1841

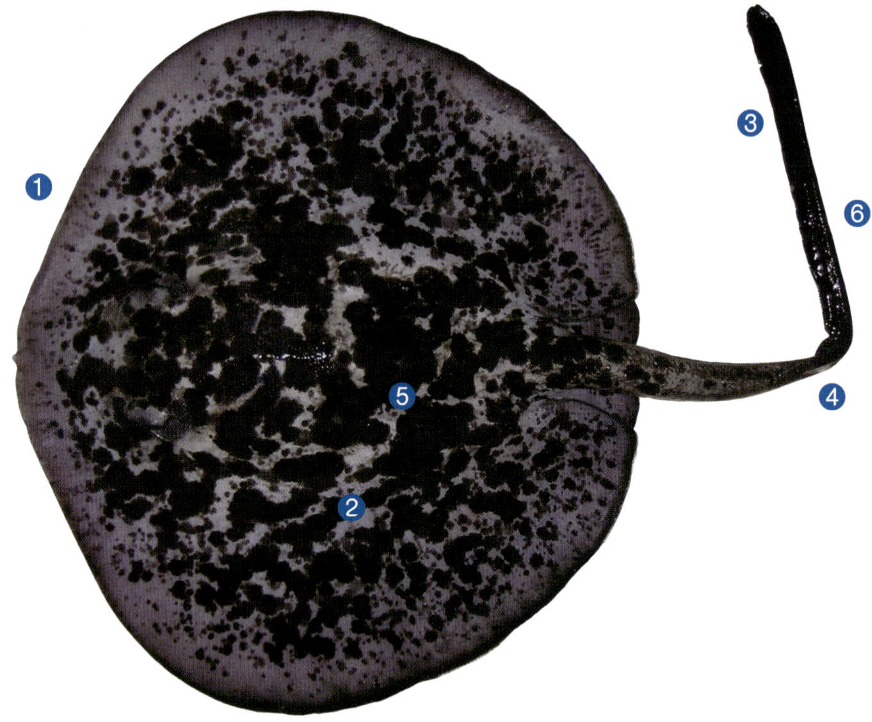

<table>
<tr><td rowspan="1">SIZE</td><td>
Maximum: 180 cm DW (330 cm TL)
Maturity: males at 100–110 cm DW; females unknown
At birth: 30–35 cm DW
</td></tr>
</table>

KEY FEATURES

❶ Disc circular

❷ Upper surface dark with white blotches and mottling (often faint)

❸ Ventral skin fold on tail deep, extending to tail tip

❹ Large stinging spine present on tail

❺ Dorsal surface of disc mostly smooth (denticles very small)

❻ Tail black posterior to sting

Worldwide distribution: Indo-West Pacific, from South Africa to Micronesia and Japan.

Gulf occurrence: Rare; observed on reef habitat off Kuwait.

Habitat & biology: Mainly found over coral reef habitats and on adjacent soft bottoms; from close inshore to at least 439 m depth. Feeds on bivalves, benthic crustaceans and small fishes. Viviparous, with histotrophy; producing about 7 pups per litter.

Conservation status: IUCN Red List: Vulnerable (assessed 2006).

Remarks: Often confused with the Cowtail Stingrays throughout its range, with all referred to as bull rays in some areas. The generic placement of this species is still under investigation; possibly belongs in the genus *Dasyatis*.

References: None

A Blotched Fantail Ray observed by divers off Kuwait

Image details: Dorsal: Bali, Indonesia (juvenile female 50 cm DW, 16th Mar 2005).

Arabian Butterfly Ray

Gymnura cf. *poecilura*

SIZE

Maximum: at least 94 cm DW

Maturity: males at 48 cm DW; female size at maturity unknown

At birth: unknown

KEY FEATURES

❶ Tail long, its length about equal to snout-vent length (when undamaged)

❷ No dorsal fin; a small stinging spine sometimes present

❸ Spiracles without a tentacle present on their posterior margins

❹ Dorsal surface typically greenish grey with faint white spots

❺ Tail banded, usually with 8 or 9 broad blackish bands

Worldwide distribution: Poorly defined; possibly restricted to the Western Indian Ocean, from the Red Sea to the Gulf and possibly Pakistan.

Gulf occurrence: Widespread and common.

Habitat & biology: A coastal species found on soft substrates. Diet unknown but probably feeds mainly on small to medium sized bony fishes like other similar-sized butterfly ray species. Biology largely unknown; probably viviparous with histotrophy like other butterfly ray species.

Conservation status: IUCN Red List: Not evaluated.

Remarks: Commonly occurs in bycatch, but of little value and often dumped. As with other butterfly rays, the colour pattern can be highly variable. Very similar to the Longtail Butterfly Ray *Gymnura poecilura* which occurs from India westwards to southern Japan but molecular analyses suggest it is likely a separate species; a taxonomic revision of this genus is currently underway. The name *Gymnura hormosensis* was provided for this species from the Gulf and the Gulf of Oman, but this not an available name as it was assigned with uncertainty and no type specimen was designated.

References: Vossoughi & Vosoughi (1999); Moore *et al.* (2012a)

An adult male Arabian Butterfly Ray landed at Sharq fish market,
Kuwait, 2011; highlights the variable colour patterns in this species

Image details: Dorsal: Kuwait Bay, Kuwait (juvenile male ~35 cm DW, 21st Aug 1985).

Longheaded Eagle Ray

Aetobatus flagellum (Bloch & Schneider, 1801)

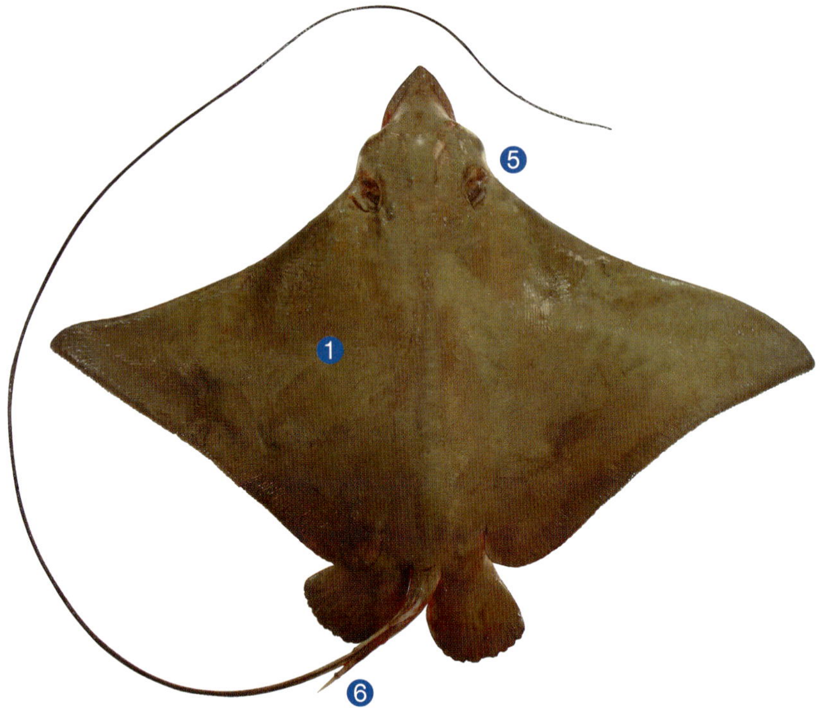

<table>
<tr><td>SIZE</td><td>

Maximum: about 90 cm DW

Maturity: males at ~50 cm DW in the Gulf; females by at least 74 cm DW

At birth: unknown; smallest free-swimming individual was 23 cm DW

</td></tr>
<tr><td>KEY FEATURES</td><td>

❶ Dorsal surface uniformly brownish, without spots

❷ Fleshy snout relatively long (very long in adult males)

❸ Nasal curtain with a very deep notch

❹ Teeth in both jaws in a single row

❺ Pectoral fin joining head at level of eye

❻ Stinging spine present behind dorsal fin

</td></tr>
</table>

Worldwide distribution: Indo-West Pacific Ocean, from the Gulf to Indonesia and Malaysia.

Gulf occurrence: Apparently restricted to estuarine-influenced areas such as the northwestern Gulf; can be common where it occurs.

Habitat & biology: One of the most common elasmobranch species in shallow, turbid waters of northern Kuwait, but apparently absent from other Gulf locations; may be naturally uncommon and mostly restricted to areas with a strong estuarine influence. Diet may include oysters, which are abundant in parts of northern Kuwait. Viviparous, with histotrophy; litter size unknown.

Conservation status: IUCN Red List: Endangered (assessed 2006).

Remarks: Previously considered to also occur in Japan where it is seasonally culled in large numbers in Ariake Bay due to heavy predation on farmed shellfish stocks; recent taxonomic investigation revealed the western North Pacific species is a distinct and much larger species, *Aetobatus narutobiei*.

References: White & Moore (2013); White *et al.* (2013); Bishop *et al.* (in press)

Myliobatidae (Eagle Rays)

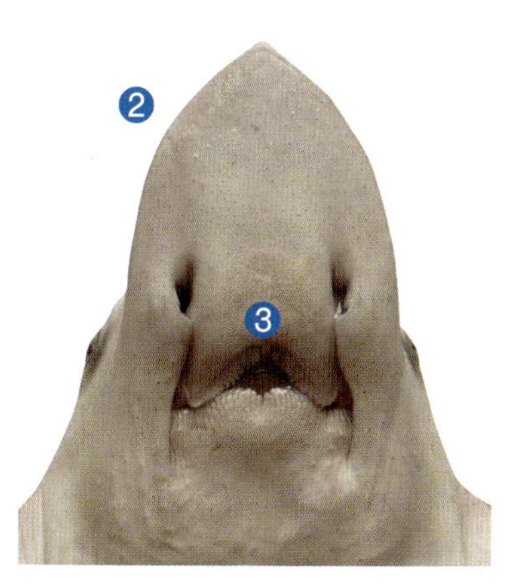

Ventral head

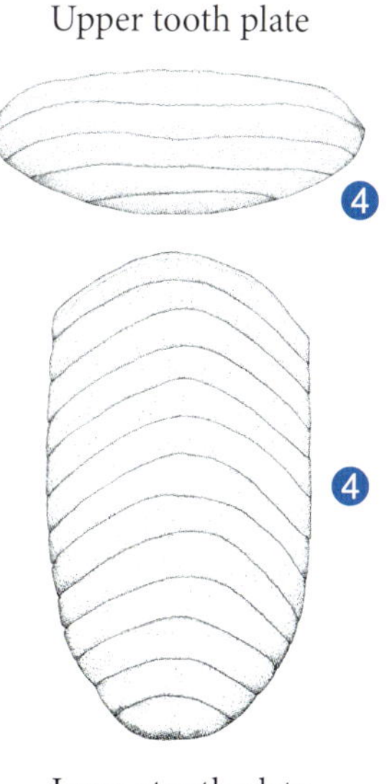

Upper tooth plate

Lower tooth plate

Image details: Dorsal: Kuwait (female ~50 cm DW, 10th Apr 2008); Ventral head: Kuwait (juvenile male 30 cm DW, 5th Apr 2011).

Whitespotted Eagle Ray

Aetobatus cf. *ocellatus*

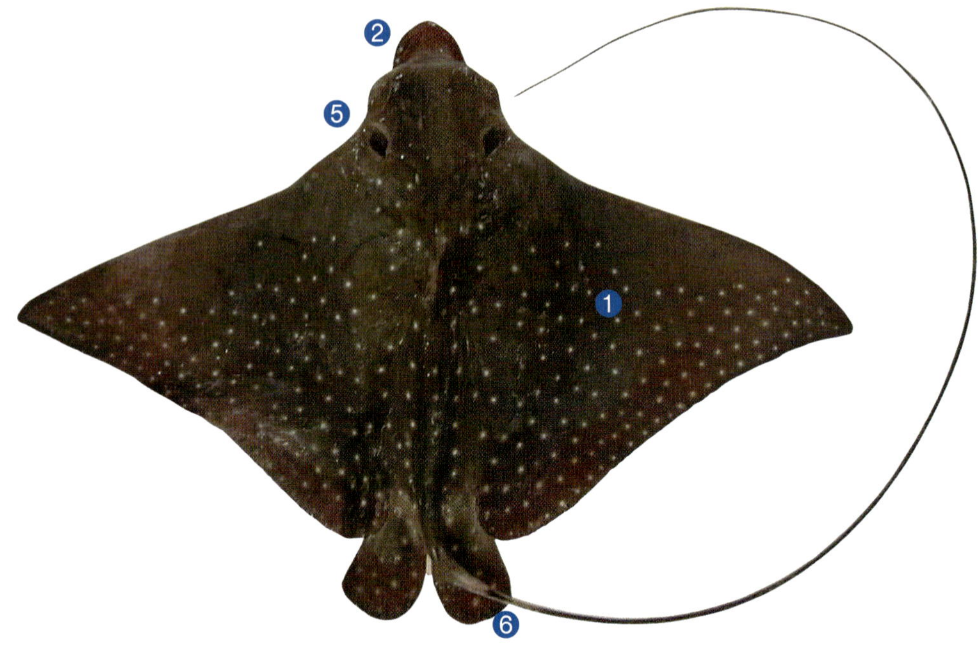

<table>
<tr><td rowspan="3">SIZE</td><td>

Maximum: 214 cm DW (155 cm DW in the Gulf)

Maturity: males by 119 cm DW in the Gulf; female size at maturity unknown

At birth: at about 18–26 cm DW

</td></tr>
</table>

KEY FEATURES

❶ Dorsal surface greenish with many small white or blue spots

❷ Fleshy snout relatively long (somewhat variable)

❸ Nasal curtain with a very deep median notch

❹ Teeth in both jaws in a single row

❺ Pectoral fin joining head at level of eye

❻ Stinging spine(s) present behind dorsal fin

Worldwide distribution: Poorly defined; possibly restricted to the northwestern Indian Ocean.

Gulf occurrence: Widespread; not common, but frequently occurs as single individuals in fish markets.

Habitat & biology: Commonly found inshore and possibly also in estuaries, but likely also occurs more offshore; often seen by divers around reefs. Diet probably consists mainly of shellfish, such as oysters, which it crushes with its strong plate-like teeth. Viviparous, with histotrophy; litter size unknown.

Conservation status: IUCN Red List: Not evaluated.

Remarks: Part of the circumtropical *A. narinari* species complex. Specimens from the Gulf are closely related to, and possibly conspecific with *Aetobatus ocellatus* which occurs throughout most of the Indo-West Pacific; molecular data has indicated that there may be a complex of species occurring in the Western Indian Ocean; further taxonomic work is required. Some specimens with numerous white occelli rather than spots.

References: White *et al.* (2010); Naylor *et al.* (2012)

Underwater image of a specimen from Qarou Island, Kuwait, 2011

Image details: Dorsal: Sharq fish market, Kuwait (juvenile male ~60 cm DW, 12th Apr 2008).

Ocellate Eagle Ray

Aetomylaeus milvus (Müller & Henle, 1841)

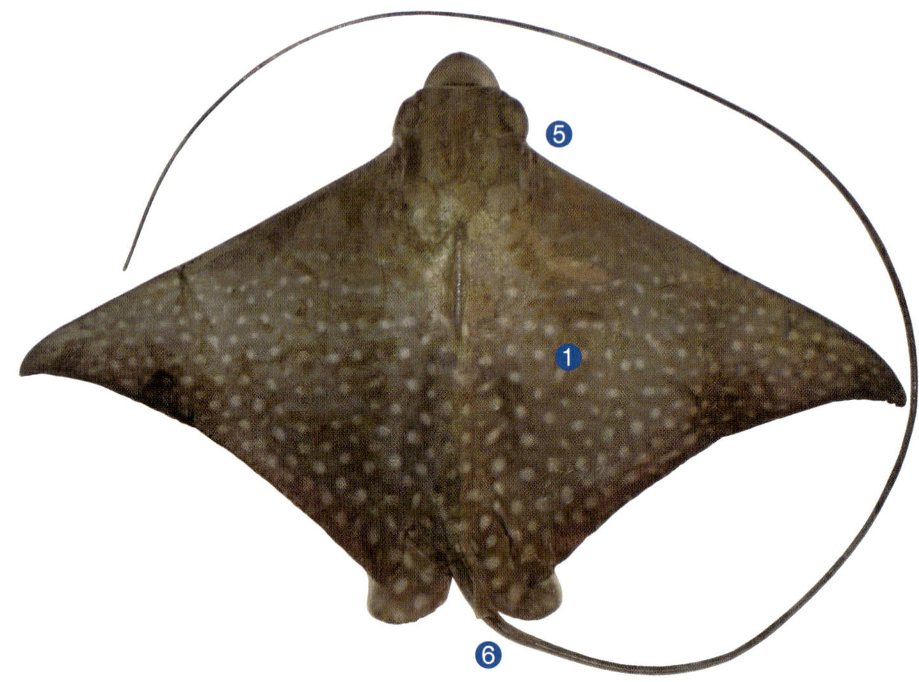

SIZE

Maximum: 123 cm DW

Maturity: males by 69 cm DW in the Gulf; female size unknown

At birth: unknown

KEY FEATURES

❶ Dorsal surface greenish brown with white spots mainly on the rear half of disc (sometimes forming lines centrally)

❷ Fleshy snout relatively short

❸ Edge of nasal curtain straight, not notched

❹ Teeth in both jaws usually in 7 hexagonal-shaped rows

❺ Pectoral fin joining head below level of eye

❻ No stinging spine on tail

Worldwide distribution: Not well defined; possibly restricted to the Western Indian Ocean; confirmed from off the Red Sea, Oman and the Gulf, but possibly also to India.

Gulf occurrence: Widespread, but not common.

Habitat & biology: Very little of the habitat or biology of this species is known; probably mainly an inshore species occurring on soft substrates. Diet probably consists mainly of benthic invertebrates. Viviparous, with histotrophy.

Conservation status: IUCN Red List: Not evaluated.

Remarks: Previously confused with the Mottled Eagle Ray *Aetomylaeus maculatus* from the Eastern Indo-West Pacific to which it is morphologically very similar; presence of ocelli and whitish lines has been thought to be a key character but both species can exhibit this colour morph. Taxonomic investigation of this species complex is currently underway. Larger specimens tend to have a small band of denticles on the bony ridge behind the head which may be a species-specific character.

References: White (2014)

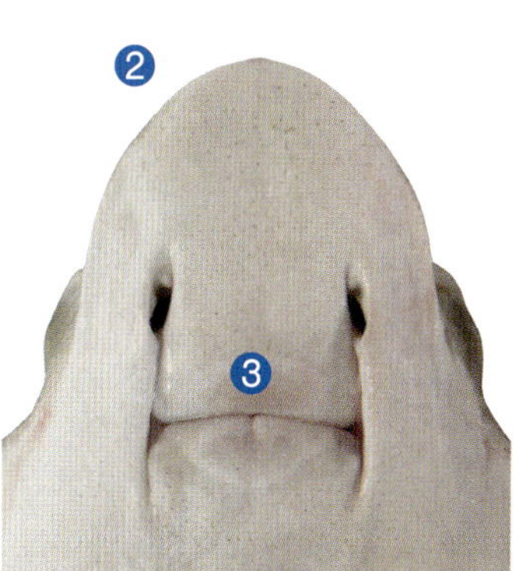

Ventral head

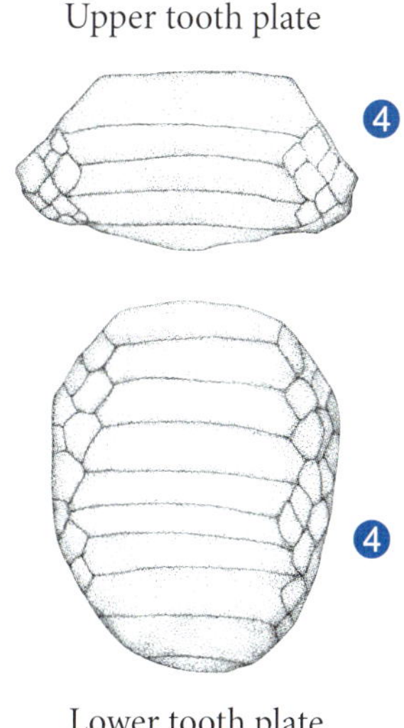

Upper tooth plate

Lower tooth plate

Image details: Dorsal: Doha, Qatar (female 55.5 cm DW, 23rd Apr 2009); Ventral head: Dubai, United Arab Emirates (female ~30 cm DW, 6th Oct 2012).

Banded Eagle Ray

Aetomylaeus nichofii (Bloch & Schneider, 1801)

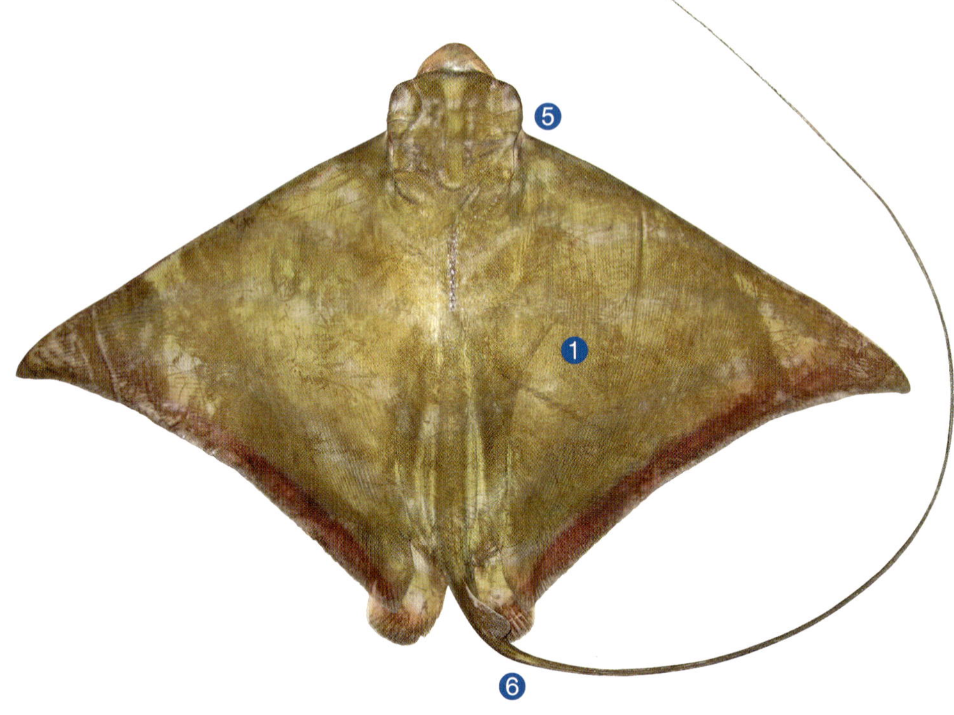

SIZE	Maximum: 72 cm DW Maturity: males at 40–42 cm DW in the Gulf; female size at maturity unknown At birth: at about 17 cm DW

KEY FEATURES	❶ Dorsal surface greyish brown with a series of broad bluish bands (most obvious in juveniles) ❷ Fleshy snout relatively short ❸ Edge of nasal curtain straight, not notched ❹ Teeth in both jaws usually in 7 hexagonal-shaped rows ❺ Pectoral fin joining head below level of eye ❻ No stinging spine on tail

Worldwide distribution: Indo-West Pacific, from the Gulf westwards to Indonesia and north to southern Japan.

Gulf occurrence: Widespread and often abundant.

Habitat & biology: An inshore species found on soft substrates to depths of at least 115 m. Diet probably consists of bivalves and small invertebrates. Viviparous, with histotrophy; produces up to 4 pups per litter.

Conservation status: IUCN Red List: Vulnerable (assessed 2003).

Remarks: Australian and New Guinea populations refer to a separate species. Molecular data has shown that specimens from the Gulf are slightly divergent from Indonesian, Malaysian and Taiwanese specimens, but this may represent population structure; further taxonomic work is required. Adult males develop bony protrusions ('horns') above the front of each orbit.

References: Moore *et al.* (2012a)

Myliobatidae (Eagle Rays)

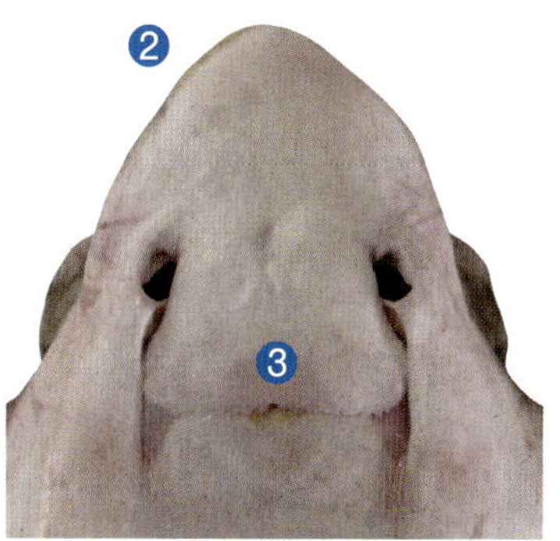

Ventral head

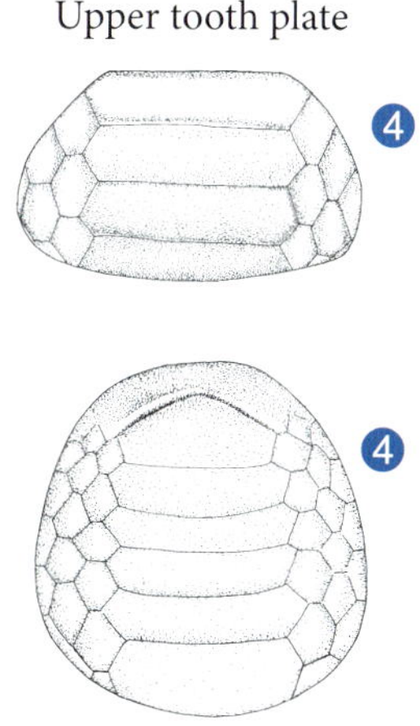

Upper tooth plate

Lower tooth plate

Image details: Dorsal and ventral head: Dubai, United Arab Emirates (female ~40 cm DW, 6th Oct 2012)

Short-tail Cownose Ray

Rhinoptera jayakari Boulenger, 1895

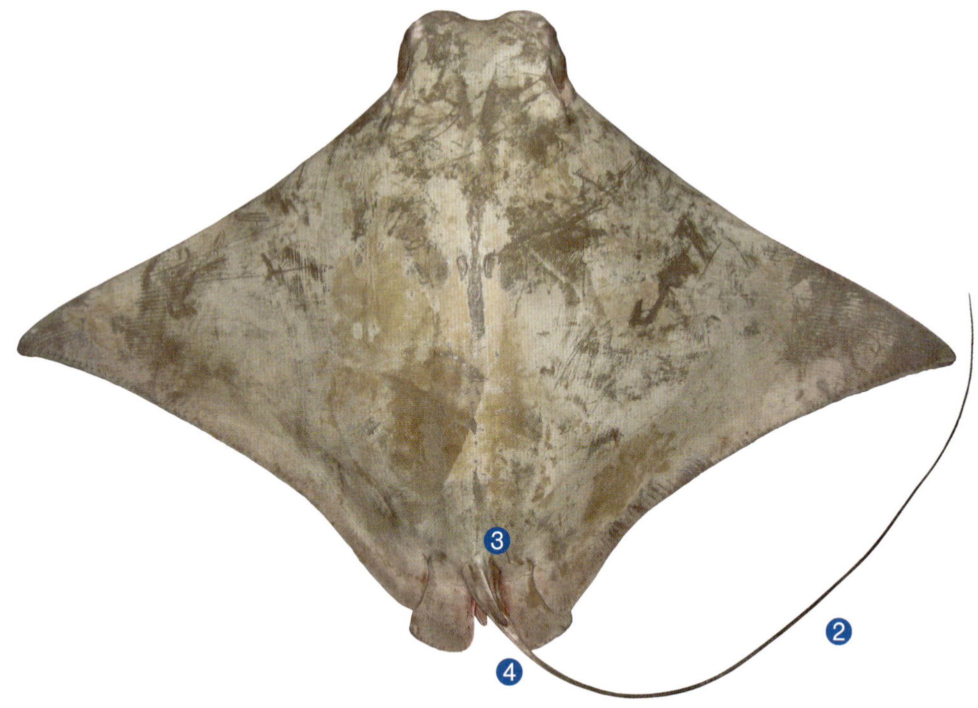

<table>
<tr><td rowspan="1">SIZE</td><td>

Maximum: 110 cm DW

Maturity: male holotype was mature at 73 cm DW; in the Gulf, females mature at about 80 cm DW

At birth: unknown; largest embryo observed was 28.5 cm DW

</td></tr>
</table>

KEY FEATURES

❶ Snout notched anteriorly to form two fleshy lobes

❷ Tail relatively short, less than width of disc

❸ Dorsal-fin origin about level with pectoral-fin insertions

❹ Stinging spine present behind dorsal fin (when undamaged)

❺ Edge of nasal curtain straight, strongly fringed

❻ Teeth plate-like, median rows widest

Worldwide distribution: Not well defined; appears to be widespread in the tropical Indo–West Pacific, from southern Africa to eastern Indonesia, north to Japan.

Gulf occurrence: Widespread and common.

Habitat & biology: Habitat and biology poorly known; pelagic, found in large schools, near the coast and well offshore. Diet unknown but probably dominated by benthic invertebrates, and possibly small pelagic fishes. Presumably viviparous, with histotrophy, as with other cownose rays.

Conservation status: IUCN Red List: Not evaluated.

Remarks: The taxonomy of this genus of rays is complicated and in urgent need of a full revision. This species has previously been confused with the Javanese Cownose Ray *Rhinoptera javanica* which has a longer tail and possibly a narrower head. Most records of the Javanese Cownose Ray may in fact refer to the Short-tail Cownose Ray.

References: Last *et al.* (2010); Moore et al. (2012a)

Rhinopteridae (Cownose Rays)

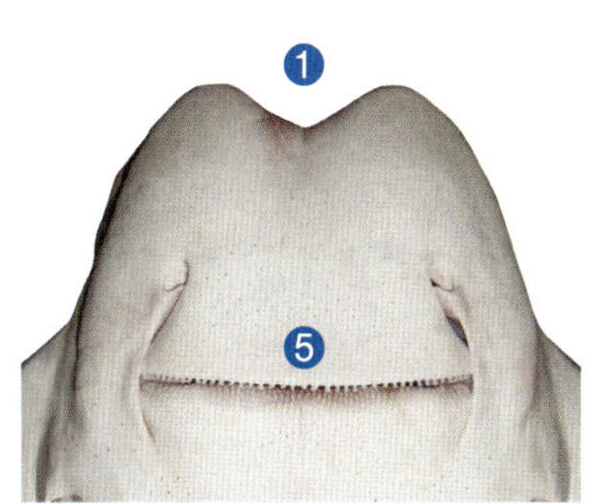

Ventral head

Cownose rays can form a large part of the unwanted bycatch in gillnet fisheries (Sharq market quayside, Kuwait, 6th Apr 2011)

Image details: Dorsal and ventral head: Dubai, United Arab Emirates (juvenile male 55 cm DW, 6th Oct 2012).

Shortfin Devilray

Mobula kuhlii (Müller & Henle, 1841)

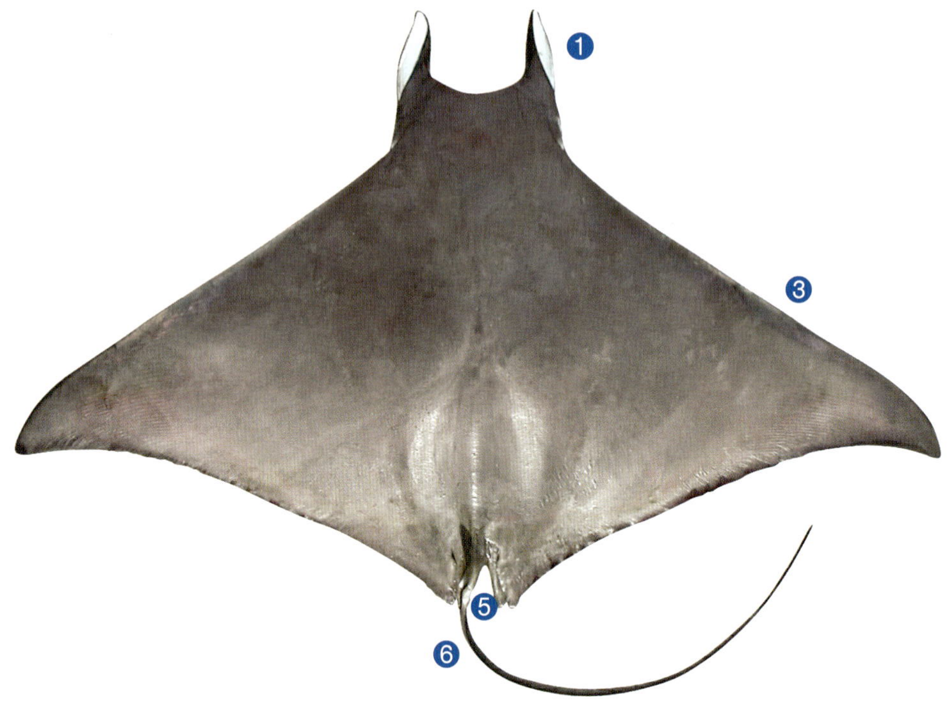

SIZE

Maximum: 135 cm DW

Maturity: both sexes by about 100 cm DW in the Gulf

At birth: unknown; largest embryo observed 31 cm DW

KEY FEATURES

❶ Head with two cephalic lobes (length variable) projecting forward

❷ Mouth subterminal (not at front of head)

❸ Anterior margin of pectoral fins straight to slightly convex

❹ Spiracle circular, located below pectoral-fin origin

❺ Dorsal fin usually with a white tip (sometimes plain)

❻ No stinging spine behind dorsal fin

Worldwide distribution: Tropical Indo-West Pacific; from South Africa to the Philippines.

Gulf occurrence: Poorly known, with the only confirmed records from Qatar; possibly widespread.

Habitat & biology: Pelagic in coastal and continental shelf waters. Diet probably consists of planktonic organisms and small fishes; the stomach of an adult male specimen recorded from Qatar contained anchovies. Viviparous with histotrophy; only a single pup per litter after an unknown gestation period.

Conservation status: IUCN Red List: Data Deficient (assessed 2009).

Remarks: Records of small devilrays with long cephalic lobes have previously been referred to as *Mobula eregoodootenke* but recent taxonomic investigation has shown that the cephalic lobes can vary in length and do not appear to be a good diagnostic feature for this species.

References: Couturier *et al.* (2012); Moore (2012)

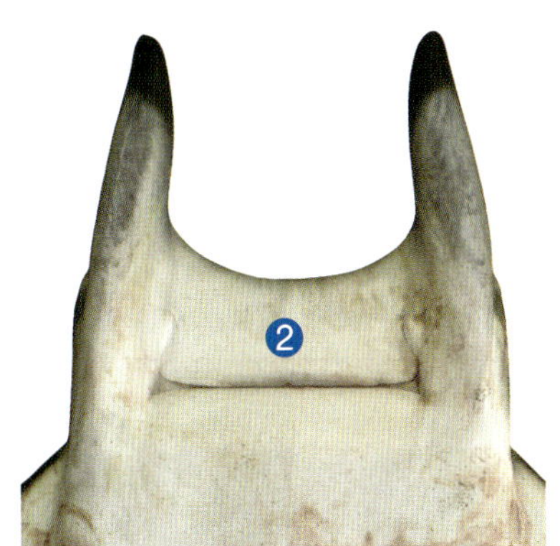

Ventral head

Dorsolateral head

Image details: Dorsal: Sarawak, Malaysia (juvenile male 90.5 cm DW, 1st Jun 2002); ventral and dorsolateral head: Al Khor, Qatar (adult male 105 cm DW, 24th Apr 2009).

Pelagic Thresher

Alopias pelagicus Nakamura, 1935

Size: Attains 390 cm TL; males mature at 247–276 cm TL and females at 264–290 cm TL; born at 130–160 cm TL.

Key features:
- ❶ Upper lobe of caudal fin nearly as long as rest of body
- ❷ No deep grooves on nape
- ❸ Eye relatively large, almost central on side of head
- ❹ First dorsal fin closer to pectoral-fin rear tip than pelvic fin

Worldwide distribution: Tropical and subtropical waters of the Indo-Pacific.

Habitat & biology: A pelagic and mostly oceanic species that occasionally enters shallower waters near outer reef edges or seamounts; from the surface to at least 152 m depth. Feeds on squids and small pelagic fishes, using its long whip-like tail to strike and stun its schooling prey. Ovoviviparous, with oophagy; two pups per litter with one in each uterus; no reproductive seasonality.

Conservation status: IUCN Red List: Vulnerable (assessed 2009).

Remarks: May occur in deeper parts of the Gulf. Occurs in Omani fisheries and may be transported overland to Gulf markets, such as Dubai.

References: Liu *et al.* (1999); White (2007)

Image details: Lateral: Dubai, United Arab Emirates (juvenile male ~180 cm TL, 26th Apr 2012).

Bigeye Thresher

Alopias superciliosus Lowe, 1841

Size: Attains 484 cm TL; males mature at 270–300 cm TL and females at 330–360 cm TL; born at 100-140 cm TL.

Key features:
- ❶ Upper lobe of caudal fin nearly as long as rest of body
- ❷ Deep groove on each side of nape
- ❸ Eye very large, upper margin almost level with dorsal surface of head
- ❹ First dorsal fin closer to pelvic fin than pectoral-fin rear tip

Worldwide distribution: Circumglobal in all tropical and warm temperate seas.

Habitat & biology: A pelagic and mostly oceanic species occasionally enters shallower waters near outer reef edges and seamounts; from the surface to at least 700 m depth; during the day found mostly in deeper waters (300–500 m) and at night in shallower waters (0–100 m). Feeds mainly on cephalopods and small to medium pelagic fishes, using its long upper caudal lobe to stun its schooling prey. Ovoviviparous, with oophagy; usually two (rarely four) pups per litter; no reproductive seasonality.

Conservation status: IUCN Red List: Vulnerable (assessed 2009).

Remarks: May occur in deeper parts of the Gulf. Occurs in Omani fisheries and may be transported overland to Gulf markets, such as Dubai.

References: Chen *et al.* (1997); White (2007)

Image details: Lateral: Java, Indonesia (juvenile female 153 cm TL, 20th Aug 2002).

White Shark

Carcharodon carcharias (Linnaeus, 1758)

Size: Attains 600 cm TL; males mature at 350–410 cm TL and females at 400–500 cm TL; born at 109–165 cm TL.

Key features:
- ❶ Caudal fin crescent shaped, with upper and lower lobes similar in length
- ❷ Large, triangular, serrated, blade-like teeth
- ❸ A strong keel on each side of caudal peduncle
- ❹ Snout long and conical

Worldwide distribution: Circumglobal in all oceans, but mostly in temperate seas.

Habitat & biology: Occurs mostly on the continental shelf, but also well offshore; found in a variety of areas from close inshore to 1280 m depth. Prefers prey with high fat content, such as marine mammals, seabirds, turtles and tuna, as well as rays and other sharks. Ovoviviparous, with oophagy; 2–17 pups per litter; gestation period possibly 18 months with a 3 year reproductive cycle.

Conservation status: IUCN Red List: Vulnerable (assessed 2009); CITES Appendix II.

Remarks: An early record of this species from Kuwait was a misidentification of the Sandtiger Shark. Not yet confirmed from waters around Arabia, but has potential to occur anywhere within its vast range from sub-polar to tropical seas.

References: Moore *et al.* (2007)

Image details: Lateral: Australia (juvenile male 246 cm TL, 11th Oct 2006).

Shortfin Mako

Isurus oxyrinchus Rafinesque, 1810

Size: Attains at least 400 cm TL; males mature at 195–215 cm TL and females at 270–300 cm TL; born at 60–70 cm TL.

Key features:
- ❶ A large lateral keel on each side of caudal peduncle
- ❷ Teeth long, slender, smooth-edged and curved with bent tips
- ❸ Snout sharply pointed (in ventral view), white ventrally
- ❹ Pectoral fins relatively short (less than length of head)

Worldwide distribution: Circumglobal in tropical and temperate seas.

Habitat & biology: A pelagic and mostly oceanic species, occasionally seen around islands and close to land; from the surface to at least 650 m depth. Feeds mainly on cephalopods, turtles, porpoises, seabirds, smaller sharks and bony fishes such as mackerels, tunas and bonitos. Ovoviviparous, with oophagy; 4–25 (usually 10–18) pups per litter; gestation may be 15–18 months with a 3 year reproductive cycle.

Conservation status: IUCN Red List: Vulnerable (assessed 2009).

Remarks: May occur in deeper parts of the Gulf. Occurs in Omani fisheries and may be transported overland to Gulf markets, such as Dubai.

References: None

Image details: Lateral: Indonesia (female 223 cm TL, 18th Aug 2005).

Longfin Mako

Isurus paucus Guitart, 1966

Size: Attains about 430 cm TL; males mature at 190–228 cm TL and females by about 245 cm TL; born at 97–120 cm TL.

Key features:
- ❶ A large lateral keel on each side of caudal peduncle
- ❷ Teeth long, pointed, smooth edged and straight (tips not bent)
- ❸ Snout broadly pointed (in ventral view), dusky to dark ventrally
- ❹ Pectoral fins very long (about equal to length of head)

Worldwide distribution: Circumglobal in tropical and warm temperate seas, but only patchily recorded; apparently rare in the Indian Ocean.

Habitat & biology: A pelagic and mostly oceanic species; from the surface to at least 220 m; probably occurring mainly in deeper oceanic waters but limited information available. Ovoviviparous, with oophagy and uterine cannibalism; 2–8 pups per litter after an unknown gestation period; may come closer to land to pup. Feeds on pelagic fishes and cephalopods.

Conservation status: IUCN Red List: Vulnerable (assessed 2006).

Remarks: Not confirmed for the northwestern Indian Ocean, but has the potential to occur in deeper water fisheries such as those off Oman.

References: None

Image details: Lateral: Indonesia (female ~240 cm TL, 26th Oct 2008).

Bigeye Houndshark

Iago omanensis (Norman, 1939)

Size: Attains 84 cm TL; males mature at about 31 cm TL and females at about 35 cm TL in Omani waters; born around 19 cm TL.

Key features:
1. First dorsal fin-origin well forward, over pectoral-fin bases
2. Gill slits about equal to eye length
3. Second dorsal fin almost as large as first
4. Upper and lower teeth small, blade-like

Worldwide distribution: Northern Indian Ocean, from the Red Sea and Gulf of Oman to Pakistan and southwestern India; possibly also the Bay of Bengal.

Habitat & biology: A benthopelagic species on continental shelves and slopes; from 110 m to well over 1000 m depth (mostly 100–250 m in Omani waters) with males possibly inhabiting deeper waters than females. Feeds on cephalopods and bony fishes. Viviparous, with yolk-sac placenta; 1–24 pups per litter with number of pups increasing with maternal size; gives birth in spring after about a one year gestation.

Conservation status: IUCN Red List: Least Concern (assessed 2009).

Remarks: May occur in deeper parts of the Gulf; common in the much deeper Gulf of Oman. Possibly a complex of species and requires taxonomic revision.

References: Henderson *et al.* (2006; 2009)

Image details: Lateral: Oman (adult male ~34 cm TL, 11th Jul 2012).

Silvertip Shark

Carcharhinus albimarginatus (Rüppell, 1837)

Size: Attains 300 cm TL; males mature at 160–180 cm TL and females at 160–199 cm TL; born at 63–81 cm TL.

Key features:
- ❶ First dorsal, pectoral, pelvic and caudal fins with prominent white tips
- ❷ Interdorsal ridge present
- ❸ Apex of first dorsal fin pointed or narrowly rounded
- ❹ Snout tip rounded (in ventral view)

Worldwide distribution: Tropical waters of the Indo-Pacific.

Habitat & biology: A mostly benthopelagic species on continental and insular shelves, around offshore islands, coral reefs areas and reef dropoffs; from the surface to 800 m. Feeds on cephalopods, eagle rays, smaller sharks and bony fishes such as groupers, mackerel, tuna and soles. Viviparous, with yolk-sac placenta; 1–11 (usually 5 or 6) pups per litter after about a one year gestation; biannual cycle.

Conservation status: IUCN Red List: Near Threatened (assessed 2009).

Remarks: May occur rarely in landings from distant fisheries.

References: Randall (1995)

Image details: Lateral: Australia (juvenile female 79 cm TL, 27th Jul 1985).

Bignose Shark

Carcharhinus altimus (Springer, 1950)

Size: Attains 283 cm TL; males mature at 190–216 cm TL and females at about 225 cm TL; born at 60–90 cm TL.

Key features:
1. First dorsal fin relatively tall, its origin over pectoral-fin insertions
2. Interdorsal ridge present
3. Snout rounded and moderately long (in ventral view)
4. Usually 15 rows of teeth on each side of upper jaw

Worldwide distribution: Probably circumglobal in tropical and warm temperate seas, but records are patchy.

Habitat & biology: A benthopelagic species found mostly near the bottom on the edge of the continental shelf and upper slopes; from 80–430 m depth. Feeds on cephalopods and bottom-dwelling fishes such as stingrays, small sharks, croakers, lizardfishes and flatfishes. Viviparous, with yolk-sac placenta; 3–15 (usually 7) pups per litter.

Conservation status: IUCN Red List: Data Deficient (assessed 2009).

Remarks: May occur in deeper parts of the Gulf; recorded from the much deeper Gulf of Oman.

References: Henderson & Reeve (2011)

Image details: Lateral: Indonesia (juvenile male 106 cm TL, 28th Mar 2006).

Galapagos Shark

Carcharhinus galapagensis (Snodgrass & Heller, 1905)

Size: Attains up to 370 cm TL; males mature at 205–230 cm TL and females at 235–240 cm TL; born at 61–80 cm TL.

Key features:
- ❶ First dorsal-fin origin level with pectoral-fin free rear tip
- ❷ Second dorsal-fin inner margin long, its length 1.6–2.1 times its height
- ❸ Interdorsal ridge present
- ❹ Upper teeth serrated, low, broadly triangular, erect to oblique

Worldwide distribution: Circumglobal in tropical and warm temperate seas, but mainly associated with oceanic islands.

Habitat & biology: A pelagic species found mainly around oceanic islands close to reefs with clear water and strong currents; from the surface to at least 285 m depth; young prefer shallower waters. Feeds on a variety of bottom dwelling prey such as cephalopods, bony fishes and elasmobranchs; larger individuals increase their consumption of rays and smaller sharks. Viviparous, with a yolk-sac placenta; 4–16 (average 9) pups per litter.

Conservation status: IUCN Red List: Near Threatened (assessed 2003).

Remarks: A young specimen recorded from Oman suggests that this species could be present in the region.

References: Henderson *et al.* (2007)

Image details: Lateral: Lord Howe Rise, Australia (juvenile male ~88 cm TL, 23rd May 2003).

Oceanic Whitetip Shark

Carcharhinus longimanus (Poey, 1861)

Size: Attains up to 395 cm TL; males mature at 168–198 cm TL and females at 175–200 cm TL; born at 60–65 cm TL.

Key features:
- ❶ First dorsal fin and pectoral fins enlarged with very broadly rounded tips
- ❷ Fins with mottled white tips in specimens over 130 cm TL
- ❸ Interdorsal ridge present
- ❹ Snout short and broadly rounded (in ventral view)

Worldwide distribution: Circumglobal in all tropical and warm temperate seas.

Habitat & biology: A pelagic and oceanic species usually found well offshore, but can occur close to land in shallower waters where the continental shelf is narrow; from the surface to at least 152 m depth. Feeds on cephalopods, crustaceans, pelagic bony fishes, stingrays, sea turtles, birds and mammalian carrion. Viviparous, with yolk-sac placenta; 1–15 (average 6) pups per litter; gestation of about one year on a biannual cycle.

Conservation status: IUCN Red List: Vulnerable (assessed 2006).

Remarks: May occur in overland landings from distant high-seas fisheries.

References: Randall (1995)

Image details: Lateral: Al-Mukalla fishmarket, Yemen (female ~140 cm TL, 18th Mar 2013).

Ganges Shark

Glyphis gangeticus (Müller & Henle, 1839)

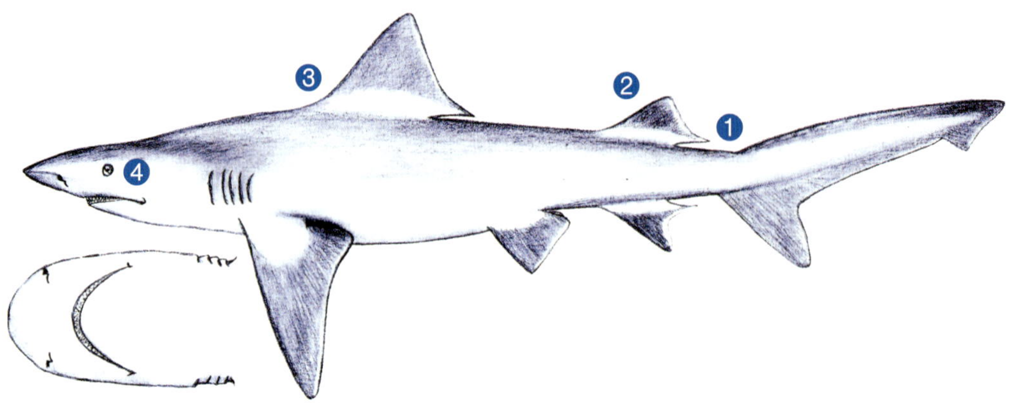

Size: Attains at least 280 cm TL, possibly larger; males mature at about 178 cm TL; born at 56–61 cm TL.

Key features:
- ❶ Precaudal pits longitudinal (not crescentic as in most whaler sharks)
- ❷ Second dorsal fin large, about half height of first dorsal fin
- ❸ First dorsal fin well forward, its origin over pectoral-fin bases
- ❹ Eyes very small

Worldwide distribution: Currently known only from the Ganges and Hooghly River systems in India, but possibly also to Pakistan.

Habitat & biology: A poorly known species which inhabits freshwater rivers and probably estuaries and inshore marine waters; probably prefers very turbid water based on its small eyes. Probably viviparous, with yolk-sac placenta as in other *Glyphis* species.

Conservation status: IUCN Red List: Critically Endangered (assessed 2007).

Remarks: Unlikely to occur, but sharks have been poorly researched in the Tigris/Euphrates system, which used to discharge close to estuaries in Pakistan before sea levels rose. *Glyphis* species are commonly misidentified as Bull Sharks.

References: Moore (2011)

Image details: Lateral: Ganges River, India (juvenile male 56 cm TL).

Blue Shark

Prionace glauca (Linnaeus, 1758)

Size: Attains 383 cm TL; males mature at 182–281 cm TL and females at about 220 cm TL; born at 35–50 cm TL.

Key features:
- ❶ Dorsal surfaces dark blue, white ventrally
- ❷ First dorsal-fin well behind pectoral-fin free rear tip
- ❸ Pectoral fins very long and scythe-like
- ❹ Snout very long and conical

Worldwide distribution: Circumglobal in all temperate and tropical oceans.

Habitat & biology: An oceanic and epipelagic species, sometimes found close to shore when the continental shelf is narrow; from the surface to about 1000 m; capable of transoceanic migrations. Feeds mainly on squids and small pelagic fishes, but also crustaceans, small sharks, cetaceans and occasionally seabirds. Viviparous, with yolk-sac placenta; 4–135 (usually 30–40) pups per litter; gestation of 9–12 months.

Conservation status: IUCN Red List: Near Threatened (assessed 2009).

Remarks: May occur in landings from offshore fisheries operating in the Arabian Sea.

References: None

Image details: Lateral: Tasmania, Australia (juvenile male 58 cm TL, 21st Jun 1996).

Spadenose Shark

Scoliodon laticaudus Müller & Henle, 1838

Size: Attains 74 cm TL; males mature at 24–36 cm TL and females at 33–35 cm TL; born at about 13–15 cm TL.

Key features:
- ❶ Head and snout strongly depressed, spade-like
- ❷ First dorsal-fin well behind pectoral-fin free rear tip
- ❸ Second dorsal fin origin well behind origin of the larger anal fin
- ❹ Pectoral fins very small, not falcate

Worldwide distribution: Western Indian Ocean, from Tanzania to at least India and Sri Lanka; possibly in the Bay of Bengal.

Habitat & biology: A coastal species which is most commonly found close to rocky bottoms and forms very large schools; from the surface to at least 13 m depth. Feeds on shrimps, crabs, cuttlefish and small bony fishes such as anchovies and codlets. Viviparous, with a unique columnar placenta; 6–18 (usually about 13) pups per litter; gestation of about 4 months.

Conservation status: IUCN Red List: Near Threatened (assessed 2009), but needs revising due to recent taxonomic studies revealing two species involved.

Remarks: May occur in landings from Pakistan, where it is common.

References: Martin & Devaraj (1997); White *et al.* (2010); Raje *et al.* (2012)

Image details: Lateral: Cochin, India (female 58 cm TL, 25th Jan 1980).

Whitetip Reef Shark

Triaenodon obesus (Rüppell, 1837)

Size: Attains 213 cm TL, but rarely above 160 cm TL; males mature at 104–116 cm TL and females at 105–122 cm TL; born at 52–60 cm TL.

Key features:
1. First dorsal and upper caudal fin with distinct white tips
2. Second dorsal fin nearly as large as first dorsal fin
3. Snout very short and broad with a blunt tip
4. Teeth smooth-edged with a strong cusplet on each side of main cusp

Worldwide distribution: Tropical waters of the Indo-Pacific.

Habitat & biology: A reef-associated species usually found near the bottom on coral reefs, often resting on the bottom or in caves and under ledges during the day; mostly in depths of 8–40 m depth, but one record from 330 m. Feeds on mostly on bony fishes, such as eels, snappers and parrotfishes, but also crustaceans and cephalopods; nocturnal feeder. Viviparous, with a yolk-sac placenta; 1–6 (usually 2 or 3) pups per litter; gestation at least 5 months.

Conservation status: IUCN Red List: Near Threatened (assessed 2005).

Remarks: Not yet reliably reported from the Gulf region. Occurs in the Red Sea. Possibly occurs on the reefs fringing the Straits of Hormuz.

References: Randall (1977; 1995)

Image details: Lateral: Bali, Indonesia (juvenile male 78 cm TL, 9th Apr 2001).

Winghead Shark

Eusphyra blochii (Cuvier, 1816)

Size: Attains 186 cm TL; males mature at 100–108 cm TL and females at about 120 cm TL; born at 32–47 cm TL.

Key features:
1. Head extremely broad and wing-shaped, its width about half total length
2. First dorsal-fin very tall, its origin over pectoral-fin bases
3. Midline of head with a shallow indentation
4. Upper precaudal pit a narrow longitudinal groove (not crescentic)

Worldwide distribution: Tropical Indo-West Pacific, from Iran to northern Australia and China.

Habitat & biology: A coastal species which is mainly found close inshore and in estuaries. Feeds mainly on small bony fishes, but also crustaceans and cephalopods. Viviparous, with yolk-sac placenta; 6–25 pups per litter; gestation of 8–11 months.

Conservation status: IUCN Red List: Near Threatened (assessed 2003).

Remarks: Likely to have occurred in the Gulf historically based on a record from off the Iranian coast in the Gulf of Oman in 1937. Localised depletions have been reported elsewhere and have possibly occurred in the Gulf.

References: Blegvad & Løppenthin (1944)

Image details: Lateral: Papua New Guinea (juvenile female 41 cm TL, 21st Jun 2014).

Smooth Hammerhead

Sphyrna zygaena (Linnaeus, 1758)

Ventral head

Size: Attains up to 400 cm TL; males mature at 210–250 cm TL and females at 220–265 cm TL; born at 50–61 cm TL.

Key features:
- ❶ Head broad, its width less than a third total length
- ❷ Anterior margin of head curved and without a central notch
- ❸ Second dorsal fin small with long rear tip, posterior margin weakly concave
- ❹ Upper precaudal pit crescentic

Worldwide distribution: Circumglobal in all tropical and warm temperate seas.

Habitat & biology: A pelagic and semi-oceanic species, sometimes occurring in coastal waters including bays and estuaries; from the surface to at least 60 m. Feeds on a variety of cephalopods, crustaceans, bony fishes, stingrays and small sharks. Viviparous, with a yolk-sac placenta; 20–50 pups per litter; gestation of 10–11 months.

Conservation status: IUCN Red List: Vulnerable (assessed 2005).

Remarks: Juveniles can occur in huge migrating schools, and possibly make incursions into the eastern end of the Gulf at times. Recently confirmed as occurring in Omani fisheries.

References: Henderson & Reeve (2011)

Image details: Lateral and head: Dubai, United Arab Emirates (female ~130 cm TL, 7th Oct 2012).

Largetooth Sawfish

Pristis pristis (Linnaeus, 1758)

Size: Attains 750 cm TL; mature at about 240 to 300 cm TL; born at 73–90 cm TL.

Key features:
- ❶ First dorsal-fin origin well forward of pelvic-fin origins
- ❷ Ventral caudal-fin lobe very short
- ❸ Rostral saw with 14–23, evenly spaced pairs of lateral teeth
- ❹ Rostral teeth present near saw base

Worldwide distribution: Circumglobal in all tropical seas.

Habitat & biology: A benthic species found in coastal, estuarine and riverine habitats, from the intertidal zone to at least 60 m depth. Feeds on small to medium-sized fishes and benthic invertebrates. Viviparous, with yolk-sac dependency; producing litters of 1–13 pups with a gestation period of about 5 months recorded for the Lake Nicaragua population.

Conservation status: IUCN Red List: Critically Endangered (assessed 2013).

Remarks: The Indo-Pacific population previously referred to as *Pristis microdon*. Two confirmed records exist from the Arabian Peninsula and could have once been widespread in the Gulf.

References: Faria *et al.* (2013); Moore (2014)

Image details: Dorsal: northwestern Australia (juvenile ~140 cm TL, 8th Nov 2002).

Bigeye Electric Ray

Narcine oculifera Carvalho, Compagno & Mee, 2002

Size: Attains at least 35 cm TL; males mature at about 24 cm TL and females by at least 32.5 cm TL; size at birth unknown.

Key features:
- ❶ Dorsal surface yellowish white with light brown to reddish brown reticulations
- ❷ Spiracles circular and large with elevated rims
- ❸ First dorsal fin taller than second dorsal fin
- ❹ Eyes large and bulging

Worldwide distribution: Known only from the Gulf of Aden and Gulf of Oman.

Habitat & biology: A benthic species found on soft bottoms in depths of 20–152 m. Viviparous, probably with yolk-sac dependency; one pregnant female contained 3 late term embryos.

Conservation status: IUCN Red List: Data Deficient (assessed 2012).

Remarks: Known to occur in the Gulf of Oman and could possibly occur in the easternmost Gulf.

References: Carvalho *et al.* (2002a)

Image details: Dorsal: near Muscat, Gulf of Oman (adult male 32.5 cm TL, 31st Aug 1988).

Pita Skate

Raja pita Fricke & Al-Hassan, 1995

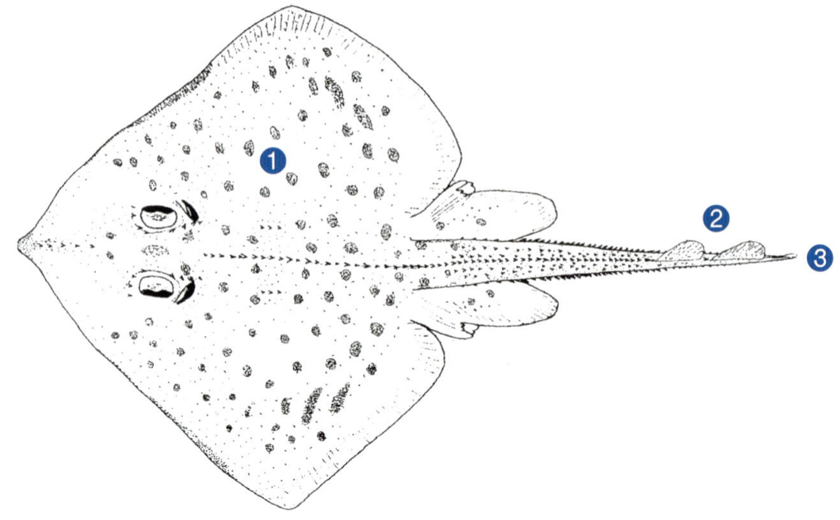

Size: Known only from one 46 cm TL female.

Key features:
- ❶ Dorsal surface with many brown spots and streaks, but no ocelli
- ❷ Two small dorsal fins of similar size near end of tail
- ❸ Small caudal fin present

Worldwide distribution: Western Indian Ocean (see remarks).

Habitat & biology: The single known specimen was reportedly collected on muddy bottoms in waters less than 15 m depth. Biology unknown, presumably oviparous, as with all other skates.

Conservation status: IUCN Red List: Critically Endangered (assessed 2009).

Remarks: Although the holotype was reportedly collected from Iraqi waters, there is some doubt as to whether this location is correct; possibly collected from waters outside of the Gulf. More specimens required to ascertain true geographic distribution of this species. This species has been placed into the genus *Okamejei*, but this is incorrect; its correct generic placement is currently being investigated.

References: Fricke & Al-Hassan (1995)

Image details: Dorsal: illustration of holotype (female 46 cm TL).

Whitespotted Whipray

Himantura gerrardi (Gray, 1851)

Size: Attains about 100 cm DW (~220 cm TL); males mature at 46–48 cm DW and females at about 64 cm DW; born at 18–21 cm DW.

Key features:
- ❶ Disc profile quadrangular, with a pointed snout
- ❷ Upper surface of disc usually with numerous white spots
- ❸ Tail long, whip-like, with alternating light and dark bands (rarely faint)

Worldwide distribution: Indo-West Pacific, but not well defined as probably a complex of species.

Habitat & biology: Occurs on soft bottom habitats, from close inshore to depths of at least 60 m. Diet unknown, but presumably consists of bivalves, crustaceans and small fishes. Viviparous, with histotrophy; gives birth to 1–4 pups per litter.

Conservation status: IUCN Red List: Vulnerable (assessed 2009).

Remarks: Probably a complex of very similar looking species, including the non-spotted Arabian Banded Whipray *Himantura randalli* (p. 110); taxonomic revision required. Known to occur in the Gulf of Oman and possibly also occurs in the Gulf.

References: Last *et al.* (2012)

Image details: Dorsal: Malaysia (adult male 62 cm DW, 3rd Jun 2002).

Mangrove Whipray

Himantura granulata (Macleay, 1883)

Size: Attains at least 141 cm DW (~350 cm TL); males mature at about 55–65 cm DW; born at 14–28 cm DW.

Key features:
- ❶ Disc almost circular in profile, snout short
- ❷ Upper surface greyish with small white flecks (often with dark mucous)
- ❸ Tail moderately long, whip-like, white behind sting

Worldwide distribution: Widespread in the Indo-West Pacific, from the Red Sea to Micronesia.

Habitat & biology: A mostly coastal species found on sandy bottoms, coral rubble and mangrove habitats, from the intertidal to depths of at least 85 m. Feeds mainly on small crustaceans. Viviparous, with histotrophy.

Conservation status: IUCN Red List: Near Threatened (assessed 2009).

Remarks: Only recently confirmed from the Western Indian Ocean (Red Sea and Gulf of Aden); apparently rare.

References: Moore (2012)

Image details: Dorsal: Indonesia (female 104 cm DW, 8th Feb 2005).

Jenkins' Whipray

Himantura jenkinsii (Annandale, 1909)

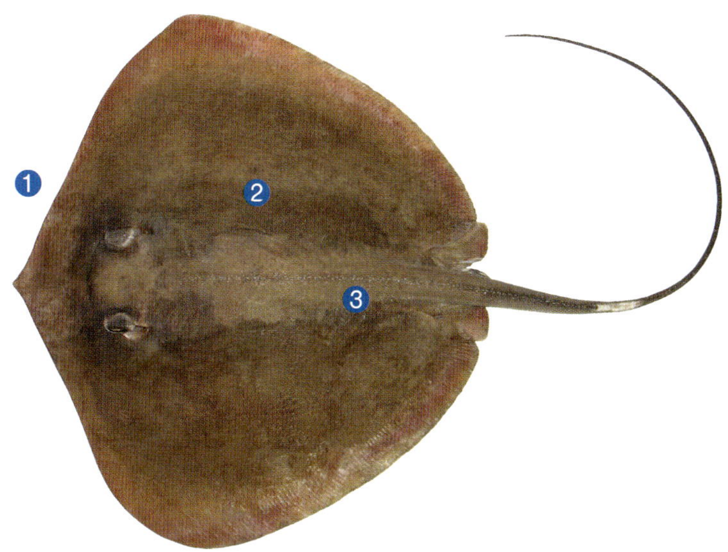

Size: Attains at 150 cm DW (300 cm TL); males mature at 75–85 cm DW; born at about 20–27 cm DW (45 cm TL).

Key features:
- ❶ Disc profile quadrangular with a short, broad snout
- ❷ Upper surface uniformly yellowish brown
- ❸ Central disc and tail with a row of upright thorns

Worldwide distribution: Tropical Indo-West Pacific, from southern Africa to northern Australia.

Habitat & biology: A coastal species found mainly on sandy bottoms and reef areas, to depths of at least 90 m. Probably feeds mainly on crustaceans and small fishes. Viviparous, with histotrophy.

Conservation status: IUCN Red List: Least Concern (assessed 2009).

Remarks: Reported as occurring in Oman, and could possibly occur in the Gulf. Distinguished from similar looking whiprays by the central row of thorns on its disc.

References: Henderson & Reeve (2011)

Image details: Dorsal: Indonesia (juvenile male 54 cm DW, 7th Dec 2005).

Bluespotted Maskray

Neotrygon kuhlii (Müller & Henle, 1841)

Size: Attains at least 47 cm DW; males mature at 22–32 cm DW and females at 24–31 cm DW; born at 11–17 cm DW.

Key features:
- ❶ Snout short with black bar (mask-like) through eyes
- ❷ Tail not whip-like, with broad black and white bands
- ❸ Large, bright, blue spots or ocelli on upper disc

Worldwide distribution: Widespread in the tropical Indo-West Pacific.

Habitat & biology: A coastal species found on sandy beds near rocky reefs, from the intertidal to depths of at least 90 m. Feeds on benthic crabs and shrimps. Viviparous, with histotrophy; producing litters of 1–3 pups per litter; Queensland populations found to have a 4 month gestation period.

Conservation status: IUCN Red List: Data Deficient (assessed 2009).

Remarks: Appears to be a complex of several species which are yet to be defined; taxonomic revision in progress. A widespread species which has been recorded off the Oman coast, which might possibly occur in the easternmost Gulf.

References: Pierce *et al.* (2009); Puckridge *et al.* (2013)

Image details: Dorsal: Oman (juvenile male 19.5 cm DW, 2012).

Bluespotted Fantail Ray

Taeniura lymma (Forsskål, 1775)

Size: Attains at least 35 cm DW (over 75 cm TL); males mature at 20–22 cm DW and females at 20–24 cm DW; female size at maturity is unknown; born at 13–14 cm DW.

Key features:
- ❶ Disc oval, with numerous bright blue spots
- ❷ Ventral skin fold on tail relatively deep, extending to tip of tail
- ❸ Tail with blue stripe on each side before sting

Worldwide distribution: Widespread in the Indo-West Pacific, from southern Africa to the Solomon Islands.

Habitat & biology: A coastal species found mainly on coral reefs and adjacent sandy flats and seagrass patches, to depths of at least 20 m. Feeds mainly on benthic molluscs, worms and crustaceans. Viviparous, with histotrophy.

Conservation status: IUCN Red List: Near Threatened (assessed 2005).

Remarks: Possibly occurs in the Gulf in coral reef habitat, but confirmed records are required.

References: None

Image details: Dorsal: Indonesia (adult female 30 cm DW, 13th Apr 2004).

Porcupine Ray

Urogymnus asperrimus (Bloch & Schneider, 1801)

Size: Attains at least 115 cm DW, possibly 147 cm DW; males mature by about 90 cm DW and females by about 100 cm DW; size at birth unknown.

Key features:
- ❶ Dorsal surface covered with long, sharp thorns
- ❷ Disc almost circular
- ❸ Tail without skin folds or stinging spines

Worldwide distribution: Indo-West Pacific, from southeastern Africa to Fiji and southern Japan; also in the tropical eastern Atlantic off central Africa.

Habitat & biology: A coastal species found on sand and mud bottoms near reefs and mangrove stands. Feeds on benthic crustaceans and bivalves; forages deeper into the substrate than most other stingrays. Viviparous, with histotrophy.

Conservation status: IUCN Red List: Vulnerable (assessed 2005).

Remarks: Probably occurs in the Gulf area since within its known range and suitable habitat is available, but confirmed records are required.

References: None

Image details: Dorsal: Marshall Islands (female 79 cm disc length, 12th Jul 1981).

Reef Manta Ray

Manta alfredi (Krefft, 1868)

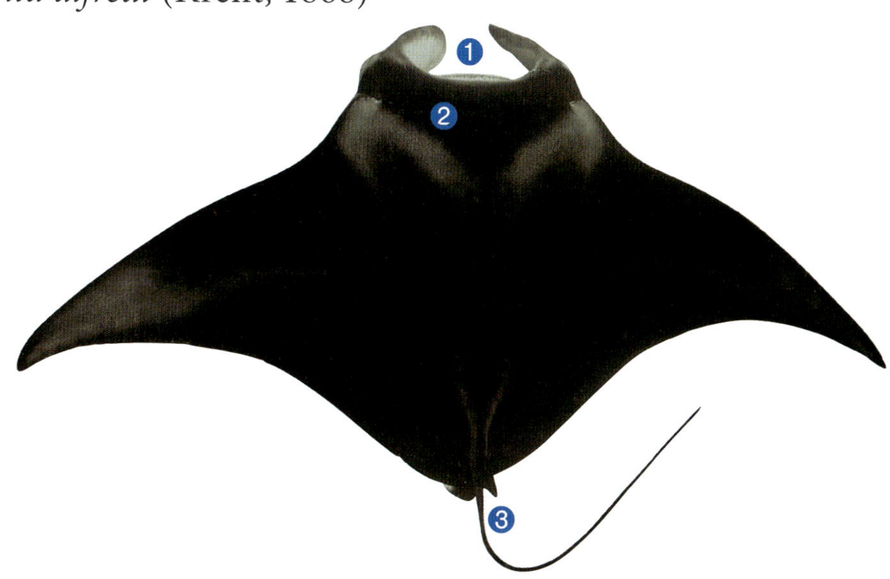

Size: Attains about 550 cm DW; males mature at about 300 cm DW and females at 390 cm DW in the southwest Indian Ocean; born at 130–150 cm DW.

Key features:
- ❶ Mouth at front of head (terminal)
- ❷ Anterior margins of white shoulder markings angling backwards, not parallel with front of head
- ❸ No calcified mass present behind dorsal fin

Worldwide distribution: Widespread in the Indian Ocean, Western and Central Pacific and the tropical Eastern Atlantic.

Habitat & biology: A coastal pelagic species, found around coral and rocky reefs, and seamounts. Viviparous, with histotrophy; a single pup per litter after a 12–13 month gestation period. Feeds on planktonic organisms and probably small bony fishes.

Conservation status: IUCN Red List: Vulnerable (assessed 2011).

Remarks: Occurs throughout the Indian Ocean so incursions into the Gulf possible.

References: Marshall *et al.* (2010); Couturier *et al.* (2012)

Image details: Dorsal: Queensland, Australia (adult).

Giant Manta Ray

Manta birostris (Walbaum, 1792)

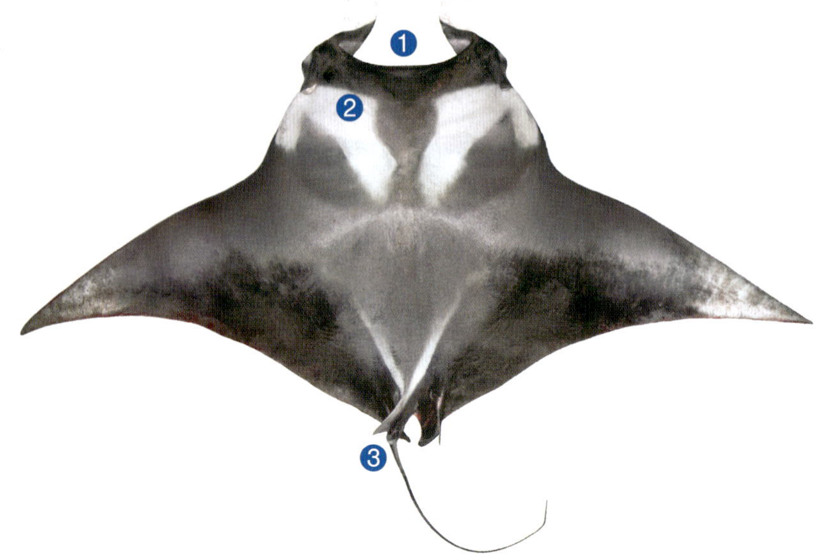

Size: Attains at least 700 cm DW (possibly 910 cm DW); males mature at about 375–400 cm DW and females at 410–470 cm DW; size at birth unknown.

Key features:
1. Mouth at front of head (terminal)
2. Anterior margins of white shoulder markings running parallel with front of head
3. A calcified mass encasing a small caudal spine present behind dorsal fin

Worldwide distribution: Circumglobal in all tropical and warm temperate seas.

Habitat & biology: A pelagic species, most often sighted along coastlines with regular upwellings, oceanic islands and offshore seamounts. Viviparous, with histotrophy; only a single pup per litter after an unknown gestation period. Feeds primarily on planktonic organisms and small bony fishes.

Conservation status: IUCN Red List: Vulnerable (assessed 2011).

Remarks: Occurs throughout the Indian Ocean and recorded off Pakistan, thus incursions into the Gulf possible.

References: Marshall *et al.* (2010); Couturier *et al.* (2012); Moore (2012)

Image details: Dorsal: Indonesia (female 338 cm DW, 26th Oct 2004).

Acknowledgements

The publication of this book has only been made possible by the generous support of a group of Kuwaiti ladies, who wish to remain anonymous, that passionately believe in the development of high-quality educational materials specific to the region. We are greatly thankful for their support, and we hope this end product repays their trust in us.

We are also thankful to the Kuwait Foundation for the Advancement of Science (KFAS), who supported one of the authors (DA) in their research, the fieldwork for which has provided many of the photographs used in this book. People that we have to thank are too numerous to mention all by name, but we would like to thank Dr Faiza Alyamani and Dr James Bishop (Kuwait Institute for Scientific Research), Major General Mohammed Al Sabah (Kuwait Coast Guard), Professor Ralf Schuster (Central Veterinary Research laboratory, UAE), Dr Haidar Murad (Public Authority for Agriculture and Fisheries, Kuwait), Dr Mehdi Jaaffar (Environmental Society of Oman), Edwin Grandcourt and Stanley Hartmann (Environment Agency Abu Dhabi), Jonathan Ali Khan (Shark Quest Arabia), Mr. Khalid Alsayegh (United Fisheries of Kuwait), Richard Peirce (Shark Conservation Society, UK), Dr. Peter Last (CSIRO) for facilitating survey work and/or providing valuable information that greatly contributed in putting this book together. One of us (WW) would also like to acknowledge the support of the CSIRO National Research Collections of Australia – the National Fish Collection, and the Oceans & Atmosphere Flagship for their support. Projects supported by the National Science Foundation (e.g. Jaws and Backbone: Chondrichthyan Phylogeny and a Spine for the Vertebrate Tree of Life; DEB-01132229) and the Australian Centre for International Agricultural Research (projects FIS/2003/037 and FIS/2006/142) have contributed greatly to the taxonomy of sharks and rays in the Indo-West Pacific as well as allowing collection of images which were used in this guide.

We are most grateful to Yaslam Saeed, Muhammed Yusur, Uzma Ali Feroze, Mr. Lateef, Ali Alhafez, and Sayed Hashim Alsayed for their valuable guidance at local landing sites and in fish markets. We are particularly grateful to all the fishermen and market traders who were happy to chat with us, answer our questions, and patiently allow us to sample their merchandise or catch. Finally, we are grateful to all those who kindly contributed photographs for use in this book.

Image copyright information

Aris Vidalis (Kuwait Turtle Conservation Project)

Aetobatus cf. *ocellatus* – underwater (p. 125); *Himantura imbricata* – underwater (p. 109).

Alec Moore

Sharks: *Carcharhinus amblyrhynchoides* – lateral, ventral head (p. 44), comparison (p. 45); *Carcharhinus amboinensis* – landed specimen (p. 49); *Carcharhinus dussumieri* – lateral, ventral head (p. 52); *Carcharhinus leiodon* – lateral, ventral head (p. 58); *Carcharhinus leucas* – lateral, ventral head (p. 60), landing (p. 61); *Carcharhinus melanopterus* – lateral, ventral head (p. 66); *Carcharhinus sorrah* – in fish market (p. 71); *Chaenogaleus macrostoma* – lateral, ventral head (p. 38); *Chiloscyllium arabicum* – ventral head (p. 26), in fish market (p. 27); *Mustelus mosis* – litter (p. 37); *Paragaleus randalli* – stomach contents (p. 43); *Rhizoprionodon oligolinx* – ventral head (p. 80).

Rays: *Aetobatus flagellum* – dorsal (p. 122), ventral head (p. 123); *Aetobatus* cf. *ocellatus* – dorsal (p. 124); *Aetomylaeus milvus* – dorsal (p. 126); *Glaucostegus granulatus* – in fish market (p. 97); *Glaucostegus halavi* – in fish market (p. 99); *Gymnura* cf. *poecilura* – dorsal (p. 121); *Himantura fai* – dorsal (p. 106); *Himantura randalli* – dorsal (p. 110), oronasal, juvenile (p. 111); *Himantura uarnak* – adult dorsal (p. 113); *Mobula kuhlii* – lateral head (p. 133); *Pastinachus sephen* – landed specimen (p. 117); *Pristis zijsron* – historic landing (p. 89, from a photo-board at the National Museum of Bahrain); *Rhina ancylostoma* – landed specimen (p. 91); *Rhinoptera jayakari* – landed specimen (p. 131); *Rhynchobatus djiddensis* – dorsal (p. 92), dorsal head (p. 93).

Carley Bansemer

Carcharias taurus – lateral (p. 34).

Charles Kilgour

Rhynchobatus djiddensis – landed specimen (p. 93).

CSIRO, Australian National Fish Collection

Sharks: *Carcharhinus amboinensis* – lateral, ventral head (p. 48); *Carcharhinus amblyrhynchos* – ventral head (p. 46); *Carcharhinus brevipinna* – adult lateral, juvenile lateral, ventral head (p. 50); *Carcharhinus falciformis* – lateral, ventral head (p. 54); *Carcharhinus humani* – ventral head (p. 56); *Carcharhinus leiodon* – holotype (p. 59); *Carcharhinus macloti* – lateral, ventral head (p. 64);

Carcharhinus plumbeus – ventral head (p. 68); *Carcharhinus sorrah* – lateral, ventral head (p. 70); *Galeocerdo cuvier* – lateral, ventral head (p. 72); *Hemipristis elongata* – lateral, ventral head (p. 40); *Loxodon macrorhinus* – lateral, ventral head (p. 74), eye (p. 75); *Mustelus mosis* – lateral, ventral head (p. 36); *Nebrius ferrugineus* – lateral, ventral head (p. 28); *Paragaleus randalli* – lateral, ventral head, (p. 42), upper and lower jaws (p. 43); *Rhincodon typus* – lateral (p. 32); *Sphyrna lewini* – in fish market (p. 83); *Stegostoma fasciatum* – ventral head (p. 30).

Rays: *Aetomylaeus milvus* – ventral head (p. 127); *Aetomylaeus nichofii* – dorsal (p. 128), ventral head (p. 129); *Anoxypristis cuspidata* – dorsal, ventral head (p. 86); *Himantura fai* – ventral head (p. 107); *Mobula kuhlii* – dorsal (p. 132), ventral head (p. 133); *Pastinachus ater* – dorsal (p. 114); *Pristis zijsron* – ventral head (p. 88); *Rhina ancylostoma* – lateral head (p. 91); *Rhinobatos* sp. – dorsal (p. 100), ventral head, interorbital space (p. 101); *Rhinoptera jayakari* – dorsal (p. 130), ventral head (p. 131); *Rhynchobatus laevis* – adult dorsal, juvenile dorsal (p. 94), ventral head, meat in market (p. 95); *Taeniurops meyeni* – dorsal (p. 118).

Possibly occurring sharks: *Alopias superciliosus* – lateral (p. 135); *Carcharodon carcharias* – lateral (p. 136); *Carcharhinus albimarginatus* – lateral (p. 140); *Carcharhinus altimus* – lateral (p. 141); *Carcharhinus galapagensis* – lateral (p. 142); *Eusphyra blochii* – lateral (p. 148); *Isurus oxyrinchus* – lateral (p. 137); *Isurus paucus* – lateral (p. 138); *Iago omanensis* – lateral (p. 139); *Prionace glauca* – lateral (p. 145); *Sphyrna zygaena* – lateral, ventral head (p. 149); *Triaenodon obesus* – lateral (p. 147).

Possibly occurring rays: *Himantura gerrardi* – dorsal (p. 153); *Himantura granulata* – dorsal (p. 154); *Himantura jenkinsii* – dorsal (p. 155); *Manta birostris* – dorsal (p. 160); *Narcine oculifera* – dorsal (p. 151); *Neotrygon kuhlii* – dorsal (p. 156); *Pristis pristis* – dorsal (p. 150); *Taeniura lymma* – dorsal (p. 157).

Daniel van Duinkerken

Dasyatis microps – dorsal (p. 104, manipulated), underwater (p. 105).

Dareen Almojil

Sharks: *Carcharhinus limbatus* – lateral, ventral head (p. 62), underwater (p. 63); *Carcharhinus plumbeus* – lateral (p. 68); *Negaprion acutidens* – lateral, ventral head (p. 76); *Rhizoprionodon acutus* – lateral, ventral head (p. 78), second dorsal/ anal fins (p. 79); *Sphyrna lewini* – lateral, ventral head (p. 82); *Sphyrna mokarran* – lateral, ventral head (p. 84); *Stegostoma fasciatum* – adult lateral (p. 30).

Rays: *Pastinachus ater* – underwater (p. 115); *Rhina ancylostoma* – dorsal (p. 90), ventral head (p. 91); *Taeniurops meyeni* – underwater (p. 119).

Possibly occurring sharks: *Alopias pelagicus* – lateral (p. 134); *Carcharhinus longimanus* – lateral (p. 143).

Government of Bahrain

Galeocerdo cuvier – landed specimen (p. 73).

John E. Randall

Sharks: *Carcharhinus amblyrhynchos* – lateral (p. 46); *Carcharhinus humani* – lateral (p. 56); *Chiloscyllium arabicum* – lateral (p. 26); *Stegostoma fasciatum* – juvenile lateral, juvenile dorsal (p. 30).

Rays: *Glaucostegus granulatus* – dorsal (p. 96); *Glaucostegus halavi* – dorsal (p. 98); *Gymnura* cf. *poecilura* – dorsal (p. 120); *Himantura fai* – underwater (p. 107); *Himantura imbricata* – dorsal (p. 108); *Himantura uarnak* – dorsal (p. 112); *Pastinachus sephen* – dorsal (p. 116); *Pristis zijsron* – dorsal (p. 88); *Torpedo panthera/sinuspersica* – dorsal (p. 102).

Possibly occurring sharks: *Scoliodon laticaudus* – lateral (p. 146).

Possibly occurring rays: *Urogymnus asperrimus* – dorsal (p. 158).

Kirsten Jensen & Janine Caira (elasmobranchs.tapewormdb.uconn.edu)

Rhizoprionodon oligolinx – lateral (p. 80).

Leonard J.V. Compagno (artwork)

Glyphis gangeticus – lateral, ventral head (p. 144).

Lydie Couturier

Manta alfredi – dorsal (p. 159, manipulated).

Mohammed Al Kanderi

Chiloscyllium arabicum – underwater (p. 27).

Matthew Pember

Stegostoma fasciatum – surface swimming (p. 31).

Richard Peirce

Carcharias taurus – dried museum specimen (p. 35); *Sphyrna mokarran* – in fish market (p. 85).

Ronald Fricke (artwork)

Raja pita – dorsal (p. 152).

Simone Caprodossi

Rhincodon typus – underwater (p. 33).

Stephen Wilson

Carcharhinus melanopterus – swimming (p. 67).

References

Abercrombie, D.L., Clarke, S.C. & Shivji, M.S. (2005) Global-scale genetic identification of hammerhead sharks: application to assessment of the international fin trade and law enforcement. *Conservation Genetics* 6, 775–788.

Assadi, H. (2001) Some aspects of reproduction biology of white cheek shark (*Carcharhinus dussumieri*) in Hormozgan waters (Oman Sea). *Iranian Scientific Fisheries Journal* 19, 1–18.

Ben Souissi, J., Golani, D., Mejri, H., Ben Salem, M. & Capapé, C. (2007) First confirmed record of the Halave's Guitarfish, *Rhinobatos halavi* (Forsskal, 1775) (Chondrichthyes: Rhinobatidae) in the Mediterranean Sea with a description of a case of albinism in elasmobranchs. *Cahiers de Biologie Marine* 48, 67–75.

Bishop, J.M., Moore, A.B.M., Alsaffar, A. & Ghaffar, A.R.A. (in press) The distribution, diversity and abundance of elasmobranch fishes in a modified subtropical estuarine system in Kuwait. *Journal of Applied Ichthyology*, in press.

Blegvad, H. & Løppenthin, B. (1944) *Fishes of the Iranian Gulf. Danish Scientific Investigations in Iran. Part III.* Einar Munksgaard, Copenhagen, 247 p.

Carrubba, R.W. & Bowers, J.Z. (1982) Kaempfer's first report of the torpedo fish of the Persian Gulf in the late Seventeenth Century. *Journal of the History of Biology* 15(2), 263–274.

Carpenter, K.E. & Niem, V.H. (eds) (1999) *FAO Species Identification Guide for Fishery Purposes. The living marine resources of the Western Central Pacific. Volume 3. Batoid fishes, chimaeras and bony fishes part 1 (Elopidae to Linophyrnidae).* pp 1397–2068. FAO, Rome.

Carvalho, M.R. de, Compagno, L.J.V. & Mee, J.K.L. (2002a) *Narcine oculifera*: a new species of electric ray from the Gulfs of Oman and Aden (Chondrichthyes: Torpediniformes: Narcinidae). *Copeia* 2002(1), 137–145.

Carvalho, M.R. de, Stehmann, M.F.W. & Manilo, L.G. (2002b) *Torpedo adenensis*, a new species of electric ray from the Gulf of Aden, with comments on nominal species of *Torpedo* from the western Indian Ocean, Arabian Sea, and adjacent areas (Chondrichthyes: Torpediniformes: Narcinidae). *American Museum Novitates* 3369, 1–34.

Chen, C.T., Liu, K.M. & Chang, Y.C. (1997) Reproductive biology of the bigeye thresher shark, *Alopias superciliosus* (Lowe, 1839) (Chondrichthyes, Alopiidae), in the northwestern Pacific. *Ichthyological Research* 44, 227–235.

Chin, A., Simpfendorfer, C., Tobin, A. & Heupel, M. (2013) Validated age, growth and reproductive biology of *Carcharhinus melanopterus*, a widely distributed and exploited reef shark. *Marine and Freshwater Research* 64, 965–975.

Cliff, G. & Dudley, S.F.J. (1991) Sharks caught in the protective gill nets off Natal, South Africa. 4. The bull shark *Carcharhinus leucas* Valenciennes. *South African Journal of Marine Science* 10, 253–270.

Cliff, G., Dudley, S.F.J. & Davis, B. (1988) Sharks caught in the protective gill nets off Natal, South Africa. 1. The sandbar shark *Carcharhinus plumbeus* (Nardo). *South African Journal of Marine Science* 7, 255–265.

Coad, B.W. & Al-Hassan, L.A.J. (1989) Freshwater shark attacks in Basrah, Iraq. *Zoology in the Middle East* 3, 51.

Coad, B.W. & Papahn, F. (1988) Shark attacks in the rivers of southern Iran. *Environmental Biology of Fishes* 23(1–2), 131–134.

Couturier, L.I.E., Marshall, A.D., Jaine, F.R.A., Kashiwagi, T., Pierce, S.J., Townsend, K.A., Weeks, S.J., Bennett, M.B. & Richardson, A.J. (2012) Biology, ecology and conservation of the Mobulidae. *Journal of Fish Biology* 80, 1075–1119.

Delshad, S.T., Mousavi, S.A., Islami, H.R. & Pazira, A. (2012) Mercury concentration of the whitecheek shark, *Carcharhinus dussumieri* (Elasmobranchii, Chondrichthyes), and its relation with length and sex. *Pan-American Journal of Aquatic Sciences* 7(3), 135–142.

Devadoss, P. & Batcha, H. (1995) Some observations on the rare bow-mouth guitar fish *Rhina ancylostoma. Marine Fisheries Information Service, Technical and Extension Series* 138, 10–11.

Dudgeon, C.L. & White, W.T. (2012) First record of potential Batesian mimicry in an elasmobranch: juvenile zebra sharks mimic banded sea snakes? *Marine and Freshwater Research* 63, 545–551.

Dudley, S.F.J. & Cavanagh, R.D. (2006) *Rhynchobatus djiddensis.* The IUCN Red List of Threatened Species. Version 2014.2. Available at: www.iucnredlist.org. Accessed 1 September 2014.

Dudley, S.F.J. & Cliff, G. (1993) Sharks caught in the protective gill nets off Natal, South Africa. 7. The blacktip shark *Carcharhinus limbatus* (Valenciennes). *South African Journal of Marine Science* 13, 237–254.

Ebert, D.A., Fowler, S. & Compagno, L.J.V. (2013) *Sharks of the World, a fully illustrated guide.* Wild Nature Press, Plymouth.

Euzen, O. (1987) Food habits and diet composition of some fish of Kuwait. *Kuwait Bulletin of Marine Science* 9, 65–85.

Faria, V.V., McDavitt, M.T., Charvet, P., Wiley, T.R., Simpfendorfer, C.A. & Naylor, G.J.P. (2013) Species delineation and global population structure of Critically

Endangered sawfishes (Pristidae). *Zoological Journal of the Linnaean Society* 167, 136–164.

Fourmanoir, P. (1976) *Requins de Nouvelle-Caledonie.* Nature Calledonienne.

Gutteridge, A.N., Bennett, M.B., Huveneers, C. & Tibbetts, I.R. (2011) Assessing the overlap between the diet of a coastal shark and the surrounding prey communities in a sub-tropical embayment. *Journal of Fish Biology* 78, 1405–1422.

Gutteridge, A.N., Huveneers, C., Marshall, L.J., Tibbetts, I.R. & Bennett, M.B. (2013) Life-history traits of a small-bodied coastal shark. *Marine and Freshwater Research* 64, 54–65.

Hall, N.G., Bartron, C., White, W.T., Dharmadi & Potter, I.C. (2012) Biology of the silky shark *Carcharhinus falciformis* (Carcharhinidae) in the eastern Indian Ocean, including an approach to estimating age when timing of parturition is not well defined. *Journal of Fish Biology* 80, 1320–1341.

Harry, A.V., Macbeth, W.G., Gutteridge, A.N. & Simpfendorfer, C.A. (2011) The life histories of endangered hammerhead sharks (Carcharhiniformes, Sphyrnidae) from the east coast of Australia. *Journal of Fish Biology* 78, 2026–2051.

Harry, A.V., Simpfendorfer, C.A. & Tobin, A.J. (2010) Improving age, growth, and maturity estimates for aseasonally reproducing chondrichthyans. *Fisheries Research* 106, 393–403.

Harry, A.V., Tobin, A.J. & Simpfendorfer, C.A. (2013) Age, growth and reproductive biology of the spot-tail shark, *Carcharhinus sorrah,* and the Australian blacktip shark, *C. tilstoni,* from the Great Barrier Reef World Heritage Area, north-eastern Australia. *Marine and Freshwater Research* 64, 277–293.

Henderson, A.C., McIlwain, J.L., Al-Oufi, H.S. & Al-Sheile, S. (2007) The Sultanate of Oman shark fishery: species composition, seasonality and diversity. *Fisheries Research* 86, 159–168.

Henderson, A.C., McIlwain, J.L., Al-Oufi, H.S., Al-Sheile, S. & Al-Abri, N. (2009) Size distributions and sex ratios of sharks caught by Oman's artisanal fishery. *African Journal of Marine Science* 31, 233–239.

Henderson, A.C., McIlwain, J.L., Al-Oufi, H.S. & Ambu-Ali, A. (2006) Reproductive biology of the milk shark *Rhizoprionodon acutus* and the bigeye houndshark *Iago omanensis* in the coastal waters of Oman. *Journal of Fish Biology* 68, 1662–1678.

Jabado, R.W., Al Ghais, S.M., Hamza, W., Henderson, A.C. & Ahmad, M.A. (2013) First record of the sand tiger shark, *Carcharias taurus,* from United Arab Emirates waters. *Marine Biodiversity Records* 6(e27), 1–4.

Jabado, R.W., Al Ghais, S.M., Hamza, W., Shivji, M.S. & Henderson, A.C. (2014) Shark diversity in the Arabian/Persian Gulf higher than previously thought: insights based on species composition of shark landings in the United Arab Emirates. *Marine Biodiversity* DOI 10.1007/s12526-014-0275-7.

Kuronuma, K. & Abe, Y. (1972) *Fishes of Kuwait*. Kuwait Institute for Scientific Research. 123 pp.

Last, P.R. & Manjaji-Matsumoto, B.M. (2010) Description of a new stingray, *Pastinachus gracilicaudus* sp. nov. (Elasmobranchii: Myliobatiformes), based on material from the Indo–Malay Archipelago, pp. 115–127. *In*: P.R. Last, W.T. White, J.J. Pogonoski (eds). Descriptions of New Sharks and Rays from Borneo. *CSIRO Marine and Atmospheric Research Paper 032*.

Last, P.R. & Stevens, J.D. (2009) *Sharks and Rays of Australia, 2nd Edition*. CSIRO, Melbourne.

Last, P.R. & White, W.T. (2013) Two new stingrays (Chondrichthyes: Dasyatidae) from the eastern Indonesian Archipelago. *Zootaxa* 3722, 1–21.

Last, P.R., White, W.T., Caira, J.N., Dharmadi, Fahmi, Jensen, K., Lim, A.P.K., Manjaji-Matsumoto, B.M., Naylor, G.J.P., Pogonoski, J.J., Stevens, J.D. & Yearsley, G.K. (2010) *Sharks and Rays of Borneo*. CSIRO Publishing, 298 pp.

Last, P.R., Manjaji-Matsumoto, B.M. & Moore, A.B.M. (2012) *Himantura randalli* sp. nov., a new whipray (Myliobatoidea: Dasyatidae) from the Persian Gulf. *Zootaxa* 3327, 20–32.

Liu, K.M., Chen, C.T. Liao, T.H. & Joung, S.J. (1999) Age, growth, and reproduction of the pelagic thresher shark, *Alopias pelagicus* in the northwestern Pacific. *Copeia* 1999, 68–74.

Malek, M., Haseli, M., Mobedi, I., Ganjali, M.R. & Mackenzie, K. (2007) Parasites as heavy metal bioindicators in the shark *Carcharhinus dussumieri* from the Persian Gulf. *Parasitology* 134(7), 1053–1056.

Manjaji-Matsumoto, B.M. & Last, P.R. (2008) *Himantura leoparda* sp. nov., a new whipray (Myliobatoidei: Dasyatidae) from the Indo-Pacific, pp. 293–301. *In*: P.R. Last, W.T. White & J.J. Pogonoski (eds). Descriptions of new Australian chondrichthyans. *CSIRO Marine and Atmospheric Research Paper 022*.

Marshall, A.D., Compagno, L.J.V. & Bennett, M.B. (2009) Redescription of the genus *Manta* with resurrection of *Manta alfredi* (Krefft, 1868) (Chondrichthyes; Myliobatoidei; Mobulidae). *Zootaxa* 2301, 1–28.

Mathew, C.J. & Devaraj, M. (1997) The biology and population dynamics of the spadenose shark *Scoliodon laticaudus* in the coastal waters of Maharashtra State, India. *Indian Journal of Fisheries* 44(1), 11–27.

McAuley, R.B., Simpfendorfer, C.A., Hyndes, G.A., Allison, R.R., Chidlow, J.A., Newman, S.J. & Lenanton, R.C.J. (2006) Validated age and growth of the sandbar shark, *Carcharhinus plumbeus* (Nardo 1827) in the waters off Western Australia. *Environmental Biology of Fishes* 77, 385–400.

Moore, A.B.M. (2010) The Smalleye Stingray *Dasyatis microps* (Myliobatiformes: Dasyatidae) in the Gulf: previously unreported presence of a large, rare elasmobranch. *Zoology in the Middle East* 49, 101–103.

Moore, A.B.M. (2011) Elasmobranchs of the Persian (Arabian) Gulf: ecology, human aspects and research priorities for their improved management. *Reviews in Fish Biology and Fisheries* 22, 35–61.

Moore, A. B. M. (2012) Records of poorly known batoid fishes from the north-western Indian Ocean (Chondrichthyes: Rhynchobatidae, Rhinobatidae, Dasyatidae, Mobulidae). *African Journal of Marine Science* 34(2), 297–301.

Moore, A.B.M. (2014) A review of sawfishes (Pristidae) in the Arabian region: diversity, distribution, and functional extinction of large and historically abundant marine vertebrates. *Aquatic Conservation: Marine and Freshwater Ecosystems* DOI: 10.1002/aqc.2441.

Moore, A.B.M. & Peirce, R. (2013) Composition of elasmobranch landings in Bahrain. *African Journal of Marine Science* 35(4), 593–596.

Moore, A.B.M., Compagno, L.J.V. & Fergusson, I.K. (2007) The Persian/Arabian Gulf's sole great white shark *Carcharodon carcharias* (Lamniformes: Lamnidae) record from Kuwait: misidentification of a sandtiger shark *Carcharias taurus* (Lamniformes: Odontaspididae). *Zootaxa* 1591, 67–68.

Moore, A.B.M., White, W.T., Ward, R.D., Naylor, G.J.P. & Peirce, R. (2011) Rediscovery and redescription of the smoothtooth blacktip shark, *Carcharhinus leiodon* (Carcharhinidae), from Kuwait, with notes on its possible conservation status. *Marine and Freshwater Research* 62, 528–539.

Moore, A.B.M., McCarthy, I.D., Carvalho, G.R. & Peirce, R. (2012a) Species, sex, size and male maturity composition of previously unreported elasmobranch landings in Kuwait, Qatar and Abu Dhabi Emirate. *Journal of Fish Biology* 80, 1619–1642.

Moore, A.B.M., Ward, R.D. & Peirce, R. (2012b) Sharks of the Persian (Arabian) Gulf: a first annotated checklist (Chondrichthyes: Elasmobranchii). *Zootaxa* 3167, 1–16.

Moore, A.B.M., Almojil, D., Jabado, R.W. & White, W.T. (2013) New biological data on the rare, threatened shark *Carcharhinus leiodon* (Carcharhinidae) from the Persian Gulf and Arabian Sea. *Marine and Freshwater Research* 65, 337–332.

Nair, R.V. & Soundararajan, R. (1976) On the occurrence of the sting ray *Dasyatis* (*Dasyatis*) *microps* (Annandale) on the Madras coast and in the Gulf of Mannar. *Indian Journal of Fisheries* 23(1–2), 273–277.

Naylor, G.J.P., Caira, J.N., Jensen, K., Rosana, K.A.M., White, W.T. & Last, P.R. (2012) A DNA sequence-based approach to the identification of shark and ray species and its implications for global elasmobranch diversity and parasitology. *Bulletin of the American Natural History Museum* 367: 263 p.

Peverell, S.C. (2005) Distribution of sawfishes (Pristidae) in the Queensland Gulf of Carpentaria, Australia, with notes on sawfish ecology. *Environmental Biology of Fishes* 73, 391–402.

Pierce, S.J., White, W.T. & Marshall, A.D. (2008) New record of the smalleye stingray, *Dasyatis microps* (Myliobatiformes: Dasyatidae) from the western Indian Ocean. *Zootaxa* 1734, 65–68.

Pierce, S.J., Pardo, S.A. & Bennett, M.B. (2009) Reproduction of the blue-spotted maskray *Neotrygon kuhlii* (Myliobatoidei: Dasyatidae) in south-east Queensland, Australia. *Journal of Fish Biology* 74, 1291–1308.

Piercy, A.N., Carlson, J.K. & Passerotti, M.S. (2010) Age and growth of the great hammerhead shark, *Sphyrna mokarran*, in the north-western Atlantic Ocean and Gulf of Mexico. *Marine and Freshwater Research* 61, 992–998.

Piercy, A.N., Carlson, J. K., Sulikowski, J.A. & Burgess, G.H. (2007) Age and growth of the scalloped hammerhead, *Sphyrna lewini*, in the north-west Atlantic Ocean and Gulf of Mexico. *Marine and Freshwater Research* 58, 34–40.

Porcher, I.F. (2005) On the gestation period of the blackfin reef shark, *Carcharhinus melanopterus*, in waters off Moorea, French Polynesia. *Marine Biology* 146, 1207–1211.

Puckridge, M., Last, P.R., White, W.T. & Andreakis, N. (2013) Phylogeography of the Indo-West Pacific maskrays (Dasyatidae, *Neotrygon*): a complex example of chondrichthyan radiation in the Cenozoic. *Ecology and Evolution* 3, 217–232.

Raje, S.G. (2006) Skate fishery and some biological aspects of five species of skates off Mumbai. *Indian Journal of Fisheries* 53(4), 431–439.

Raje, S.G., Das, T. & Sundaram, S. (2012) Relationship between body size and certain breeding behavior in selected species of Elasmobranchs off Mumbai. *Journal of the Marine Biological Association of India* 54(2), 85–89.

Randall, J.E. (1977) Contribution to the biology of the whitetip reef shark (*Triaenodon obesus*). *Pacific Science* 31(2), 143–164.

Randall, J.E. (1995) *Coastal Fishes of Oman*. Crawford House Publishing, Bathurst, Australia. 439 pp.

Randall, J.E. & Compagno, L.J.V. (1995) A review of the guitarfishes of the genus *Rhinobatos* (Rajiformes: Rhinobatidae) from Oman, with description of a new species. *The Raffles Bulletin of Zoology* 43(2), 289–298.

Robinson, D.P., Baverstock, W., Al-Jaru, A., Hyland, K. & Khazanehdari, K.A. (2011) Annually recurring parthenogenesis in a zebra shark *Stegostoma fasciatum*. *Journal of Fish Biology* 79, 1376–1382.

Robinson, D.P., Jaidah, M.Y., Jabado, R.W., Lee-Brooks, K., Nour El-Din, N.M., Al Malki, A.A., Elmeer, K., McCormick, P.A., Henderson, A.C., Pierce, S.J. & Ormond, R.F.G. (2013) Whale sharks, *Rhincodon typus*, aggregate around offshore platforms in Qatari waters of the Arabian Gulf to feed on fish spawn. *PLoS One* 8(3), 1–10.

Rowat, D. & Brooks, K.S. (2012) A review of the biology, fisheries and conservation of the whale shark *Rhincodon typus*. *Journal of Fish Biology* 80, 1019–1056.

Stevens, J.D. & Lyle, J.M. (1989) Biology of three hammerhead sharks (*Eusphyra blochii, Sphyrna mokarran* and *S. lewini*) from northern Australia. *Australian Journal of Marine and Freshwater Research* 40, 129–146.

Stevens, J.D., Pillans, R.D. & Salini, J. (2005) *Conservation assessment of* Glyphis *sp. A (speartooth shark),* Glyphis *sp. C (northern river shark),* Pristis microdon *(freshwater sawfish) and* Pristis zijsron *(green sawfish).* Final Report to DEH. 84 pp.

Vaudo, J.J. & Heithaus, M.R. (2011) Dietary niche overlap in a nearshore elasmobranch mesopredator community. *Marine Ecology Progress Series* 425, 247–260.

Vossoughi, G.H. & Vosoughi, A.R. (1999) Study of batoid fishes in northern part of Hormoz Strait, with emphasis on some species new to the Persian Gulf and Sea of Oman. *Indian Journal of Fisheries* 46(3), 301–306.

Wallace, J.H. (1967) The batoid fishes of the east coast of southern Africa. Part III: skates and electric rays. *Investigational Report, Oceanographic Research Institute* 17, 1–62.

White, W.T. (2007) Biological observations on lamnoid sharks (Lamniformes) caught by fisheries in eastern Indonesia. *Journal of the Marine Biological Association of the United Kingdom* 87, 781–788.

White, W.T. (2012) A redescription of *Carcharhinus dussumieri* and *C. sealei*, with resurrection of *C. coatesi* and *C. tjutjot* as nominal species (Chondrichthyes: Carcharhinidae). *Zootaxa* 3241, 1–34.

White, W.T. (2014) A revised generic arrangement for the eagle ray family Myliobatidae, with definitions for the valid genera. *Zootaxa* 3860(2), 149–166.

White, W.T. & Harris, M. (2013) Redescription of *Paragaleus tengi* (Chen, 1963)

(Carcharhiniformes: Hemigaleidae) and first record of *Paragaleus randalli* Compagno, Krupp & Carpenter, 1996 from the western North Pacific. *Zootaxa* 3752, 172–184.

White, W.T., Last, P.R., Stevens, J.D., Yearsley, G.K., Fahmi & Dharmadi (2006) *Economically Important Sharks and Rays of Indonesia.* ACIAR Publishing, Canberra, 329 pp.

White, W.T., Bartron, C. & Potter, I.C. (2008) Catch composition and reproductive biology of *Sphyrna lewini* (Carcharhiniformes, Sphyrnidae) in Indonesian waters. *Journal of Fish Biology* 72, 1675–1689.

White, W.T., Last, P.R. & Naylor, G.J.P. (2010) *Scoliodon macrorhynchos* (Bleeker, 1852), a second species of spadenose shark from the Western Pacific (Carcharhiniformes: Carcharhinidae), pp. 61–76. *In*: P.R. Last, W.T. White, J.J. Pogonoski (eds). Descriptions of New Sharks and Rays from Borneo. *CSIRO Marine and Atmospheric Research Paper 032.*

White, W.T., Last, P.R., Naylor, G.J.P., Jensen, K. & Caira, J.N. (2010) Clarification of *Aetobatus ocellatus* (Kuhl, 1823) as a valid species, and a comparison with *Aetobatus narinari* (Euphrasen, 1790) (Rajiformes: Myliobatidae), pp. 141–164. *In*: P.R. Last, W.T. White, J.J. Pogonoski (eds). Descriptions of New Sharks and Rays from Borneo. *CSIRO Marine and Atmospheric Research Paper 032.*

White, W.T., Furumitsu, K. & Yamaguchi, A. (2013) A new species of eagle ray *Aetobatus narutobiei* from the Northwest Pacific: an example of the critical role taxonomy plays in fisheries and ecological sciences. *PLOS One* 8(12), 1–11.

White, W.T. & Moore, A.B.M (2013) Redescription of *Aetobatus flagellum* (Bloch & Schneider, 1801), an endangered eagle ray (Myliobatoidea: Myliobatidae) from the Indo-West Pacific. *Zootaxa* 3752, 199–213.

White, W.T. & Weigmann, S. (2014) *Carcharhinus humani* sp. nov., a new whaler shark (Carcharhiniformes: Carcharhinidae) from the western Indian Ocean. *Zootaxa* 3821, 71–87.

Whitney, N.M. & Crow, G.L. (2007) Reproductive biology of the tiger shark (*Galeocerdo cuvier*) in Hawaii. *Marine Biology* 151, 63–70.

Scientific names index

Common names index